Ideological Hesitancy in Spain 1700–1750

PUBLICATIONS OF THE *BULLETIN OF HISPANIC STUDIES*

Textual Research and Criticism

Textual Research and Criticism (TRAC) publishes Spanish, Portuguese and Latin-American texts of literary, linguistic or historical interest not otherwise available in modern editions. The texts are accompanied by a substantial introductory monograph and full apparatus of critical footnotes, and the series is firmly aimed at a scholarly readership. The series also publishes literary and critical studies.

Scholars are invited to apply to the Editors for further information and to submit a brief summary of their projected book. Contributions will be assessed by eminent Hispanists in the appropriate areas, and should not exceed 400 pages of typescript. The editorial address is: *Bulletin of Hispanic Studies*, Department of Hispanic Studies, The University of Liverpool, PO Box 147, Liverpool, L69 3BX.

The Editors of TRAC are indebted to the E. Allison Peers Fund, University of Liverpool, for a subvention which has generously assisted publication of this book.

PUBLICATIONS OF THE *BULLETIN OF HISPANIC STUDIES*
Textual Research and Criticism

Ideological Hesitancy in Spain 1700–1750

by I. L. McCLELLAND

LIVERPOOL UNIVERSITY PRESS

First published 1991 by
Liverpool University Press
PO Box 147, Liverpool, L69 3BX

British Library Cataloguing-in-Publication Data

McClelland, I.L.
Ideological hesitancy in Spain 1700–1750.
—(Publications of the Bulletin of Hispanic
Studies—textual research and criticism)
I. Title II. Series
946

ISBN 0–85323–097–8 Cloth
0–85323–137–0 Paper

Set in Linotron 202 Sabon by
Wilmaset, Birkenhead, Wirral, England

Printed and bound in the European Community by
Redwood Press Limited, Melksham, England

CONTENTS

To
Dr Gillian Rodger
with my gratitude
for her inspiring friendship

PREFACE

My intention in this study has not been to give a systematized account of crises in Spain's eighteenth-century thought, or to discuss eighteenth-century arguments for and against recognizable *ilustrismo*, or intellectual enlightenment. Much excellent research has already been published on the subject. Rather have I been interested to observe the naturalness of that period's atmosphere of ideological confusion. For which reason typical examples have been chosen of hesitancy and perplexity in writers of relatively reasonable and liberal views: those whose reasonable understanding of new directions of thought is partially limited by reasonably human caution. There are at least two kinds of reasonableness in this respect: that of the far-sighted genius, and that of the intelligent man who thinks from the standpoint of his present, or local, intellectual experience. The latter's influence is usually indirect, certainly less conspicuous than that of the blind traditionalist whose opposition to *ilustrismo* is, out of fear, dramatically emotional. Yet it is the rational, half-way attitude of intelligent uncertainty, especially in well-known academicians, which most radically encouraged confusion in ill-informed masses of the general public, which ironically impeded a right understanding of the period's Feijoos, Martínezs, Islas, and their compeers, and which incidentally complicated polemic.

The difference between the anti-*ilustrado*, or anti-enlightened, and the middle-way *ilustrado* is that the former is largely inspired by theoretic prejudice, while the middle-way *ilustrado*, who thinks in more practical terms, is hesitatingly cautious. The anti-*ilustrado* is academically blind. The middle-way *ilustrado* is prepared to believe that experimental Science and rationalist thinking in general might conceivably disclose truths as yet unknown. His hesitancy is derived, at least temporarily, from natural doubts and confusion over, for example, the acceptable range of accuracy in new technology. A theoretic philosopher, however openminded, could be at a disadvantage in assessing results obtained by the new scientific machinery used in eighteenth-century experimentalism. Only when the impeding *rôle* of calculated caution is taken fully into account can the complexity of any period's mental movements be intimately understood. Those who hesitate are as psychologically influential as those who knowledgeably assert or ignorantly deny. The atmosphere of an epoch cannot be judged merely by its ideological warfare: by outright assertions from its immortal geniuses, and the violent contradictions of their ill-informed opponents. Lurking suggestibility, disturbed uncertainty in an epoch's by-ways of thinking, express its three-dimensional character. My present interest is in this psychological aspect of eighteenth-century reality.

CHAPTER 1

Reason of Unreason in the Spanish *Vulgo*

The persistent opposition experienced by rationalist reformers of the Siglo de las Luces (Century of Enlightenment) is most easily, and all too readily, judged by posterity to be a result merely of blind ignorance and prejudice. Had this been true, the reformers' task would have been much easier than it was. Irrational ignorance and prejudice certainly were factors with which *ilustristas*, and *ilustrados* – that is, modern rationalists in mere spirit or in practical fact – together with enlightened thinkers of any other century, had to reckon. In 1751, the medical scientist José Ortega, who had been asked to nominate the most reputable Botanist for the Real Sitio de San Fernando, warned against national or individual pride and prejudice in making scientific appointments:

> . . . pues unos de nuestros más dañosos defectos es el infatuarnos fácilmente presumiendo que sabemos mucho, y con esta aprensión no queremos obedecer a nadie, ni reconocer por superior a quien imaginamos erradamente que no tiene qué enseñarnos.[1]
>
> (. . . for among our most dangerous defects is that of becoming obsessed with the assumption that we know a great deal, and because of this idea we want to obey nobody, and neither do we want to recognize as superior someone whom we mistakenly imagine as having nothing to teach us.)

But areas in which *ilustrismo* strove with far greater difficulty were intellectual areas of reasonable distrust – reasonable, that is, for the period and in the national circumstances. Blind prejudice did not always explain a reluctance to favour seeming threats to established forms of reasoning or practical ways of life, least of all in a country with a history of universal greatness. Both initial distrust and sustained reluctance sometimes expressed an honest, reasonable fear of revolutionary theories and techniques which seemed, at first sight, to deny apparent facts hitherto formulated doctrinally as such, or sanctified by academic, religious and social tradition. Often warfare was conducted on intelligent lines, in grey areas of understanding where the opponents of notable *ilustrado*s are characterized by genuine puzzlement and a strong sense of public responsibility. In time some might become partially converted to new theories by reflecting on new evidence.

Others remained wholly inimical. Yet outside black or white areas of plain assertion and denial, of downright insult and retaliatory sarcasm, the very activity in reasoned distrust on the part of intelligent anti-rationalists helped to familiarize the general public with varying points of view and so to present change as evolutionary.

The inner atmosphere of the century, then, was tense with apprehension about the various ways in which society could be affected practically by new principles of thinking. An obvious example was the effect of recent scientific inventions, such as the microscope, on medical debate, and consequently on the confidence of the man-in-the-street who relied on familiar forms of treatment. Equally alarming was the apparent threat to spiritual intuition by rationalist investigation of everyday religious practices; or the tendency of rationalists to enter into international preoccupations and so, apparently, to neglect or discredit national interests and to encourage foreign customs even in the most obvious guises of dress and entertainment. Such fears were those of average thinkers, so similar to average thinkers of any other century, and who constituted, explained Feijoo, whose interest in psychology was curiously objective, the *vulgo*-majority.

What Swift called the 'rabble'[2] and Feijoo the '*vulgo*' was a lively entity whose hostility to rationalistic reform both writers attempted to analyse. The descriptions 'rabble', or *vulgo*, whatever their historical connections, were not usually meant in 'enlightened' circles as references to illiterate peasant-masses or mobs, but to the average man, the general reading public, or the average scholar of professional status afflicted, so Martín Martínez said in his *Filosofía scéptica*, by the 'peste de la pasión'(plague of passion). For, as the *Resurrección del Diario* . . . later emphasized, 'también en los literatos hay Vulgo' (among literary men too there is a common herd).[3] In other words, the *vulgo* – a term that could be used in varying degrees of disparagement from tolerance to contempt – had, for the Enlightened, a meaning socially classless. It denoted the tendency in every age, and in every kind of society, intellectual or non-intellectual, aristocratic, bourgeois or peasant, to think as a society instead of recognizing, as grouped individuals, the individual potentialities of individual ideas, and risking the discomfort of subsequent uncertainties. One of the best known of Feijoo's Discourses was devoted to explaining precisely this interpretation.[4]

Not that every enlightened scholar had, himself, escaped from the preconditioning force of inherited authorities, ideas, and practices, from national or personal pride, from fear or distrust of the unexplored, or from the thrall of dogmatic assertion and slogan. Not even Feijoo, much less Swift, was immune from the polemical infection of subjective interest. But the few who could think for themselves and judge evidence dispassionately were generally able to recognize and analyse their own diseases, rectify their diagnoses of their own and other people's complaints, and balance their judgments accordingly. Since the reasonable caution behind intelligent forms of herd-mentality is a fact of folk-life in every age – for not all change advocated by free-thinkers can be proved sound, practical, healthy or morally admissible –, the reformer's task of clearing a way through herd-suspicion in order to make contact with herd-reason was formidable: as formidable then as is his task today of presenting a case for nuclear power or racial equality. The average man's dependence on block-opinion and his consequent aversion to the loneliness of individual judgment or acceptance of open verdicts, his instinct to imitate accepted attitudes, expect definitive rules and authority, and use accepted terminology, meant that he and the individual reformer spoke different languages and needed an interpreter to mediate between them. This exercise of interpretation, often unskilled and misleading, which seemed so frustratingly unsatisfactory to all parties concerned, is one of the most vital activities of eighteenth-century polemic. Its most notable feature was its inaccuracy. Those who do not understand the essentials of, for example, scientific evidence, will argue about incidental minutiæ easier to assimilate; about applications, illustrations, or unfamiliar phraseology. The *vulgo* tended and tends to argue, off the point, about irrelevant detail. Yet the most stimulating effect of interpretative polemic was that muddled manner of disturbing preconceived forms of expression. Muddlemindedness can be healthier than mental rigidity.

A different way in which the 'plague of passion' manifested itself was in the exercise of the imagination. Certainly in this respect the *vulgo* could be influenced by the unusual and attracted by innovation. For the average citizen any difference between the spiritual reality of religion and the assumptions of superstition was difficult to define. The method of exteriorizing indefinable mystery by sensory imagination was usually a therapeutic means of converting the incomprehensible into illustrative positiveness. Dramatized superstition could be comforting because explicable. It had often helped communities to accommodate themselves to prevailing states of physical alarm. More, a wilful choice of irrationality can sometimes free the mind for adventures into the unknown, or mean that, like Don Quijote, a man or community may tend to prefer what is imaginatively attractive rather than accept a state of uninspiring reality. Spaniards are an imaginative race, especially on visual planes, and to many members of the *vulgo*-classes imagination had its own internal and external logic in every aspect of life and existed assertively in its own right.[5]

Naturally, too, the *vulgo* responded to new possibilities of pleasure, new varieties of entertainment such as newspapers; an increasing sophistication of stage-machinery; the sensationalism of supposed miracle, magic, scandal or adventure, anything, in fact, which could be contained within its mental and emotional appreciation of amenity. Herd-imagination is more difficult for rationalists to deal with than herd-reason, because it is not codified by the *vulgo*'s philosophical authorities, and though the average, the *vulgo*-scholar, was not entirely immune to superstitious influences, it was not usually the *vulgo*-scholar in this regard with whom reformers had to contend. Their engagements against imaginative preoccupations were chiefly concerned with uneducated or semi-educated members of the *vulgo*: those dependent on feeling, outer appearance, the electrifying personality of a popular speaker, the drama of exciting circumstance, of coincidence, surprise, disaster and nervous disturbance.

The dialectical means of communication used by many reformers in combat with *vulgo*-scholars were unpractical for cases where the 'peste de la pasión' raged in less intellectual sections of the *vulgo*. Feijoo treated the ignorant claims of herd-imagination with uncomplicated disgust. Isla treated it with artistic ridicule. Yet less intellectual members of the *vulgo* were not necessarily moronic. Prevailing circumstances, such as lack of precise knowledge of nature's causes and effects, the naturally constructive element in human imagination, the Church's sanction for belief in invisibilities, perhaps even a more highly developed sixth sense or intuition than that characterizing rationalistic generations of the future, made herd-values of feeling and imagination seem relatively logical, not simplistically perverse or depraved. Consequently, to discredit the force of superstition, and any other effect of 'passion', herd-imagination had to be justly understood and intelligently analysed. Sometimes in polemic it happened that impassioned anti-reformists could temporarily confound the logic of reformists with the appositeness of their home-spun psychology.

Judging by appearances, the illustrious John Wesley in England could believe and assert that rain had obligingly stopped falling in deference to the holy intent of his outdoor preaching.[6] In Spain the equally illustrious Matías Marquina could relate the same miracle to the preaching of St Anthony of Padua whose outdoor congregation remained dry when rain fell all around.[7] The enlightened Dr Johnson, according to Boswell and Joshua Reynolds, who seem to have been confiding in each other about the matter, was not without a few superstitious inhibitions. Boswell instances Johnson's elaborate care to enter a doorway on one particular foot.[8] If, then, such scholars, whom nobody could classify as *vulgo*, had retained certain *vulgo*-

instincts, is it surprising that lesser men should model any form of the unknown to the shape of folk-predisposition?

Tensions in Spain, produced especially by unassimilated reports of scientific research, had their parallels everywhere else. Too often it was, and still is, assumed that in certain other countries, dominating, individual *ilustrados* expressed the open curiosity of whole nations, or at least addressed themselves to fellow citizens willing to learn from them. Inductive thinking in Europe was nowhere characteristic of the mind of Europe's *vulgo*. As the Scots medical scholar, Smollett, during his foreign travels, observed from Paris in 1763: 'It would be absurd to conclude that the Welsh or Highlanders are a gigantic people, because those mountains may have produced a few individuals near seven feet high. It would be equally absurd to suppose the French are a nation of philosophers because France has given birth to a Descartes, a Maupertuis, a Réamur, and a Buffon.'[9] The reasons for nervous fears of unfamiliar apparatus and techniques, or problems of disturbance created by assaults on authorized theory, are discussed by English authors such as Locke, Lillo, Fielding, Richardson, Johnson and Boswell, and analysed by Rousseau, Voltaire and their European disciples. War and political changes in Spain doubtless prepared the way for further change in that they produced a general atmosphere of insecurity. Incidentally, however, they brought about a reaction of commoner-nationalism in everyday affairs, and a distrustful attitude towards foreigners and foreign ways of thinking. Which at various times, and in similar circumstances, was displayed by other countries also.

The conception of inductive reasoning for all purposes disturbed most serious scholars who, like Mayáns y Siscar, were troubled about the degree to which laws of Science, as deduced by Aristotle, could, and should, be changed.[10] Indirectly it also disturbed the general public who relied on, for instance, the traditional methods of diagnosis and treatment used by physicians trained in Aristotelian philosophy. Of equal concern to serious Europeans was the relationship between Experimental Science and religious teaching. An understandable aversion to subjecting any biblical statement to profane analysis had caused the Catholic Church to delay official recognition of Copernican evidence. A passion-plague of half rational, half emotional thinking between Anglicans and Protestant reformers in England was as devious as that between Catholics and non-Catholics. The crisis over Jesuit influence in France, Portugal, and Spain occasioned polemic just as interestedly off-the-point as, in its local circumstances, it was locally relevant. Usually, in fact, the aims of reformers were temporarily obscured by debates over minutiæ.[11]

The frustrating, yet not unhealthy, spread of polemic in Spain reaching all *vulgo*-classes, intellectual or otherwise, was activated in part by the publication of academic treatises and translations of foreign treatises, and by increasing numbers of popular tracts. In part, perhaps most vitally, it was an inevitable result of the development of the popular Press which, in the first half of the eighteenth century, emerged as a refreshing stimulant. The establishment or re-conditioning of literary and scientific Academies under Bourbon influence in various parts of Spain provided obvious new sources of information and criticism.[12] The very title, not to say the contents of the self-consciously objective journal of literary criticism, *Diario de los literatos de España, en que se reducen a compendio los escritos de los Autores Españoles, y se hace juicio de sus obras, desde el año 1737 (Journals of the literary men of Spain in which the writings of Spanish authors, from the year 1737, are summarized and critically examined)*, sounded for most *ilustristas* too chilling a note. Heated polemic roused by specific causes on emotional grounds is one thing. An open predisposition to academic criticism on selective planes of disinterestedness is another. An international call to criticism for criticism's intellectual sake might in most countries appeal only to a minority. And Spain was less ready for a development of the exercise than France or Britain. The *Diario*'s calculated aspirations stated in its Introduction could seem discomforting to the average man of strong feeling and set principles. Emphasis is on impartiality:

> Entre las virtudes del siglo pasado, se veneran por más útiles a la República de las Letras la humildad de reconocer lo limitado de las fuerzas intelectuales, para la instrucción universal, a que naturalmente aspira el entendimiento humano, y la solicitud de los medios que la facilitan, felicísimamente logrados en la institución de los Diarios o Jornales.[13]
>
> (Among the virtues of the last century are venerated, as most useful to the Republic of Letters, the humility of recognizing the limitation of intellectual forces for universal instruction, to which, naturally, the human understanding aspires, and the care concerned with the means of facilitating this, [now] most happily achieved in the institution of the Journals and Periodicals.)

This idea of a journal setting itself up as a publicly governing body of criticism would take some time to establish itself in society, and public attitude had much to do with the relative shortness of the *Diario*'s life. Perhaps in Spanish circumstances the *Diario*'s title and intention could sound offensively self-righteous and opinionated. One wonders if a more imaginative and so more attractive advertisement for a Spanish equivalent to the *Spectator*, for example, might not have been the name which Pitillas applied to himself when criticizing in the *Diario* public favourites such as the dramatist Cañizares. I want to be, he said, a 'crítico Quijote' (a critical Don Quijote). Which invitingly suggests an incidental sense of humour: a quality common to the Spanish mind but better developed, in eighteenth-century criticism, in Britain, and too often lacking among Spanish *ilustristas* and *ilustrados*. One reason for much of the successful teaching in *ilustrismo* of Benito Jerónimo Feijoo was his ability to laugh, even at himself. Humour can unblock closed entrances to understanding. While the *Diario* certainly approved of Pitillas' contribution, its editorial board was possibly too dependent, for the journal's public acceptance, on the chance strains of imagination mellowing the attitudes of occasional contributors.

Elsewhere, in regions less literary and more technical, the fact that intellectual treaties were now often published in Spanish rather than Latin, produced other possibilities of concerned incomprehensions involving problems both for writers who, in Latin, would have been comprehensible to colleagues in accepted academic contexts, and for the general public of non-academic readers who could seize only on conspicuous details definable through common knowledge, and would fail to relate them to the right context, with results of worrying misunderstanding. The *Resurrección del Diario de Madrid* in 1748 echoes a commom outcry on this subject:

> Tales materias [that is, technical matters] piden el sello de la Lengua Latina, para que el Vulgo bajo no las maltrate. Constituye éste muy mal, lo que no entiende bien. Y mal puede entender lo que está tan fuera de su capacidad.[14]
>
> (Such matters require the seal of the Latin language to prevent the lower orders of the General Public from misusing them. This order constitutes badly what it does not understand well. And [indeed] it can only understand badly what is so far outside its capacity.)

But the mind of the general public was chiefly swayed, backwards and forwards, by reports in the more general and popular press, especially those concerned with sensational foreign inventions or opinions, and with Spanish interpretations of these. General speculations together with informed and uninformed polemic healthily exteriorized the confusion over changing standards, however frustratingly it delayed the conscious acceptance of inductive thinking. Therefore we do well to take honest confusion into account when trying to understand the ulterior reason for *vulgo*-unreason. General confusion eventually could lead to a greater elasticity of judgment in an increasing number of individuals. Also through one particular institution, the theatre, the mind of the general public could be re-orientated unawares. A growing tendency to criticize social conditions and publicize means of amelioration produced the highly popular propaganda drama typical of the second half of the century. In theatres the public could be familiarized with innovation through incidental comment, assumption or general mental attitude: a technique seen at its maturity in Comella's play, *El fénix de los criados o*

María Teresa de Austria, which broaches the sensitive new subject of medical inoculation.[15] During the last decades of the century propaganda drama effortlessly publicized new ideas or social suggestions and so performed a powerful function similar to that of modern radio or television. Along such circuitous paths of erratic process did the Age of Reason in all countries doubtingly move. And it is the working realities of process, rather than the finished effects of change and progress, on which the following pages concentrate.

NOTES

1 See Juan Riera, *Medicina y ciencia en la España ilustrada: Epistolario y Documentos I*, Monografías X (Valladolid: Univ. of Valladolid Press, 1981), p. 100.

2 See Swift's Preface to *Tale of a Tub* (London: Hamish Hamilton [The Modern Library], 1930), p. 388.

3 See *Filosofía scéptica. Extracto de la Física antigua y moderna* . . . (Madrid, 1730), p. 299; *Resurrección del Diario de Madrid* . . . (Madrid, 1748), Prólogo.

4 See *Teatro crítico universal* (Madrid, 1726-1740), Vol. II. 'Voz del Pueblo'.

5 See I. L. McClelland, *Benito Jerónimo Feijoo* (New York: Twayne, 1969), pp. 81ff, 87ff.

6 See Wesley's *Journal* (London: Everyman, No. 106, 1938), Vol. II: Feb. 25th, 27th, 1755 (p. 293); May 18th, 1757 (p. 375); June 26th, 1757 (p. 381); July 20th, 1757 (p. 386); etc.

7 See Marquina's *Escuela general histórica* . . . (Madrid, 1751), Vol. I, p. 182.

8 See Boswell's *Life of Samuel Johnson* (London: Everyman, 1935), Vol. I, p. 301. Boswell thought of it as '. . . some superstitious habit which he had contracted early, and from which he had never called upon his reason to disentangle him'.

9 Tobias Smollett, *Travels through France and Italy* (London: John Lehmann, 1949).

10 See Vicent Peset, *Gregori Mayans i la cultura de la il·lustració* (Barcelona: Curial, 1975), pp. 41, 239, 248, *et passim*.

11 See, for example, Peset, *op.cit.*, pp. 216ff, 225, 258, 260, 340, 343, 375, etc.

12 See I. L. McClelland, *Spanish Drama of Pathos* (Liverpool: Liverpool University Press, 1970), Vol. II, pp. 531-32, and pp. 426ff.

13 *Diario de los literatos de España* . . . Vol. I. (Madrid: Ant. Marín, 1737), Introducción.

14 *Resurrección del Diario de Madrid, o Nuevo Cordón crítico general de España* (Madrid, 1748), p. 33.

15 See the Preliminary Chapter of Francisco Aguilar Piñal's *La Real Academia Sevillana de Buenas Letras en el siglo XVIII* (Madrid: C.S.I.C., 1966), pp. 3ff.

CHAPTER 2

The False Alarm of 'Scepticism'

The early decades of Spain's eighteenth century were fallow years of mental re-adjustment involving, severally or together, curiosity, distrust, confusion, passive and active assimilation, enterprise, hesitation and unease. Typical of the well informed thinkers of the period was the man who neither wholly committed himself to ideals of free scientific inquiry, nor closed his mind entirely to the importance of scientific discovery. Therefore, to understand the manner of his preparations for the age of rationalism, one must understand the reasons for his inner reservation and the nature of his contribution to the prevailing atmosphere of tension and cross-purpose. Especially is it necessary to observe how contradictory emotions – as may be present in uneasy curiosity – could operate abrasively in any one individual, or how intellectually susceptible men, like the influential Torres Villarroel, might echo both fashionable modernism and fashionable conservatism in fashionable terminology. The popular Torres was not as positively enlightened as he and his admirers supposed. Nor was he as penetrating as they believed in his assessment of the problem entailed in interpretations of scientific evidence. Many humbler, more cautious, or more timorous thinkers often perceived deeper complexities in the spirit of the European times than Torres, with his arrogant impatience, had noticed.

It is known that the technical reasoning of Newton's Physics, which after 1687, that is, after the publication of his *Principles of Philosophy*, largely determined the form of Europe's scientific revolution, had been quietly absorbed and analysed by a few Spanish scholars now ready to associate in thought with the few international experts capable of interpreting Newton constructively. Individual Spaniards could take their place among Europeans who correctly understood Bacon's technique of openmindedness and could correctly adopt it for the reception of Newton's Physics. These, the precociously enlightened, were they who set in motion the slow, undramatic change taking place, Dr Peset has recently shown,[1] between 1687 and 1727, on Spain's outskirts of Sevilla, Cataluña, Valencia, and other grounds less hostile than the Spanish interior to foreign ways of thought. Outstandingly significant in this respect is his account of how the appointment of two modern-minded

Professors of Medicine from Valencia, Dr Antoni Garcia i Cervera and his pupil, Dr Andrés Piquer, to royal appointments in Madrid, the favour shown to Piquer by the Marqués de la Ensenada, the special concessions made to him by Fernando VI for the purchase of such foreign books as he deemed useful,[2] and his appointment as Vice President of the Real Academia Médica de Madrid at once extended knowledge of medical attitudes of the University of Valencia and roused the antagonism of leading Madrid doctors.[3] Of such stuff was the dynamism of enlightenment indeed to be made. Eventually, however, innovators are obliged to communicate with their academic Faculties, State Universities with the State, and, in the eighteenth century, with the Church. On a more conspicuous level, those individuals wishing to change the ways of society are obliged eventually to communicate with the general public. Furthermore, the nature of the force and development of new thought is determined by the way in which it is received by the thinker's colleagues and society in general, and the extent to which an innovator needs to defend himself. Polemic, for instance, can reduce him to fighting triviality with triviality. Therefore it was in this process of communication that Spanish advocates of, for example, Baconian and Newtonian experimentalism, found themselves involved in alarmist wars of words, to the detriment of bookish apologists like Martínez, and to the greater glory of literary geniuses like Feijoo.

Early eighteenth-century polemic over scientific enlightenment might be said to have started publicly with a conflagration. Much of the alarm first occasioned by Spain's most widely informed scientists was due to the use of a single word, 'sceptical', as applied to Medicine, by Martín Martínez. To the *vulgo* the word 'scepticism' was more redolent of religious heresy than of either Greek doctrine or Baconian openmindedness. Its use by Martínez in 1722, in the title of his *Medicina scéptica*,[4] meaning medical acknowledgment of medical uncertainties, might have been less notoriously provocative and alarming had not Feijoo, who liked the word as much for academic as for dramatic reasons, flourished it spiritedly in his 'Apología' to Martínez's second edition of 1727,[5] and imaginatively paraded its every possible Baconian application in his own treatises.

Martínez was a strictly academic reasoner, a Professor of Anatomy and several times President of the Regia Sociedad Médico-Chímica of Sevilla: a port important in its own right, and which, as Dr Aguilar Piñal has emphasized, was then becoming a centre of considerable intellectual activity and independence.[6] For its part the Sociedad Médico-Chímica of Sevilla was said to be discarding Scholastic presuppositions. 'Ha desterrado', asserts one of Martínez's sponsors, Professor Miguel Marcelino Boix of the University of

Alcalá, 'todas las cuestiones escolásticas médicas, que se usan, como cosa inútil, y de ningún provecho para el uso práctico'. (It has disclaimed all the scholastically medical questioning in use as valueless and of no importance for practical purposes.)[7] Martínez, counting on local support, could afford a certain tone of self confidence. It is true that his arid style must have restricted knowledge of his work to a relatively small circle. The *Medicina scéptica*, a treatise on the rights of practical researchers to mental independence, and on the practical results of modern clinical experiment, is written in the Classical form of philosophical discussion among three representatives of three respective points of view. These are Galen, the conservative; a 'Chemist' or modernist; and Hippocrates, the sceptic with whom Martínez for the most part identifies himself. We have to remember that Hippocrates' work was not known at first hand and therefore was regarded as a scientific symbol, to be variously interpreted. It could conveniently serve as a sign of anonymity and sceptical disinterest, though Martínez probably did himself a disservice by speaking under a Classical name, however medically reputable. For it fomented confusion, and allowed one of his earliest critics, López de Araujo, to argue that no evidence has ever been brought to prove that the real Hippocrates could in any sense be called a sceptic.[8] New ideas were guaranteed greater publicity by the artistic style of Feijoo. Yet neither Martínez nor Feijoo was speaking of scepticism in the abstract. One of the reasons why their declared 'scepticism' provoked alarm was because they were applying it to everyday purposes: to practical matters which the *vulgo* dangerously half understood, and on which, accordingly, the *vulgo* could form its own opinions.

If Martínez and Feijoo, who discarded academic Latin in favour of Spanish, had managed to substitute another term for 'sceptical', the *vulgo* might have had time to acclimatize itself slowly to the change in methods of thought. Still, slow assimilation is not the only means of conversion. Shock-tactics at times can be healthily stimulating, and the fury of polemic, thrusting Feijoo into notoriety, was convenient to his cause. But the resulting sense of confusion and alarm, spreading among all but the most judicious scholars, was a realistic quality of contemporary mental life, and determined the sometimes exaggerated, the over-tendentious, the over-aggressive reactions of the enlightened minority. Martínez for this reason errs on the side of academic pedantry, Feijoo on the artistic side of rudeness.

One of the censor-sponsors of the *Medicina scéptica*, Father Master Fr. Juan Interian de Ayala, a retired Professor of Theology of Salamanca, who approved of this treatise, foresaw that serious objections might arise over Martínez's apparently unorthodox terminology, and attempted to prepare the author's way by insisting that his adjective 'sceptical' referred exclusively to non-religious matters. Ayala rightly feared that the word would be misinterpreted:

> . . . que a muchos temo haga más novedad de la que trae consigo el término, el cual nada significa más que indeterminada, irresoluta y considerativa. Y estas calidades, si se considera bien, las debe tener toda humana ciencia, y natural filosofía; sino en todas, a lo menos en muchas de sus partes, y especulaciones.[9]

> (. . . which to many people, I fear, suggests something stranger than the word itself implies: for that word means nothing more than indetermination, irresolution, and thoughtful consideration. And these qualities, it should be realized, ought to apply to all human science and natural philosophy, if not in all parts and speculations at least in many of them.)

For Spain in particular the word 'sceptical' had alarmingly religious connotations. Also, in general, it was, perhaps still is, morally disconcerting and polemically disruptive to introduce any public to strange new ideas by undermining its long established habits of terminology.

This last was no purely Spanish problem. The establishment of the Age of Reason in England and France did not establish in every Englishman and Frenchman a conscious or unconscious acceptance of scientific rationalism as a guiding principle. Certainly disturbances in orthodox thought experienced in Northern countries at the time of the Reformation made newer mental disturbances easier to propagate. But the influence of a new mode of thought depends not only on the history of social circumstances, but also on the size of the minority presenting it and on its social and political backing. In eighteenth-century England the Wesley brothers were raising a storm of alarm and antagonism among intellectuals and non-intellectuals, clergy and laity, by trying to accommodate to the independent thinking of their religious revival such orthodox words as 'faith', 'justification', 'works', which the Catholic Church and the Church of England used in a different category of meaning, referred to in a Papal Decree of 1724. The mob violence experienced by religious innovators was largely due to the result of the public's being asked to use familiar words in an unfamiliar way. Here, too, one wonders if the Wesleyan message might not have had a smoother passage if the brothers had conveyed it in terms that held no traditional 'association of ideas'. These last were the words of John Locke quoted to John Wesley by a correspondent who goes on to specify:

> People long accustomed to explain the essential things of Christianity, in such a particular way, and never having observed how they can be explained in any other, transfer their zeal for these essential things to their own way of explication; and believe there is a necessary connexion between them, when in fact there is not. This has produced many mischiefs and animosities, among all sorts of people.[10]

But again, paradoxically, much of the strength of the Wesleyan movement, like the accumulating strength of Spain's *ilustrados*, was initially due to the notoriety of its terminological polemic.

Martínez's sponsor Ayala was intelligently aware of the risks of unorthodox banner-headlines. Though professing to know nothing of Medicine, as a discipline in itself, he knew, from his training in Philosophy and Theology, that to the majority of the Catholic orthodox, Science, Medicine, and other disciplines had so far been regulated under the jurisdiction of Theology. The established medical authorities were the ancient ones approved by the Church, and medical interpretation was limited by theological doctrine. Specifically Ayala had in mind the supposedly divine approval for Hippocrates and Galen assumed to be demonstrated in *Ecclesiasticus*, where we are told to honour the physician. For since the only medical Schools known to the biblical writers, it had been argued, were the Schools of Hippocrates and his interpreter Galen, these exponents of Medicine had been awarded what amounted to theological infallibility. Consequently Ayala was afraid, he said, that the strength of prejudiced dogmatism in established Schools would allow their adherents to know only what they thought they knew already: 'con que les parece que saben aquello a que están persuadidos' (so that it seems to them that they know what they have learned to know).[11]

When publicized by polemic, Martínez's 'scepticism' was seen to incorporate certain other words almost equally provocative. Conspicuous in a different *Aprobación* (Editorial Sponsorship) to the edition of 1722 is the assertion of Professor Boix of the Chair of Medicine in Alcalá, that the *Medicina scéptica* had been written to 'unteach' medical assumption: '. . . no se ha escrito tanto para enseñar cuanto para desenseñar' (. . . it has not been written so much to teach as to unteach),[12] or in Martínez's own words: 'Muy difícil es enseñar, pero mucho más difícil desenseñar' (It is very difficult to teach, but it is much more difficult to unteach);[13] and Boix reminds readers that the supposed supreme authority of Hippocrates in Medicine is open to criticism in that his work has been known to posterity only at second hand. In his own Introduction Martínez, knowing that he moved in dangerous areas, warily offered his title as the symbol of a modern need: the need of the practical scientist inspired by the Bacons, Boyles, Sydenhams, Gassendis etc., to liberate himself from the unpractical words of Aristotelian Dialectics in order to proceed to unprejudiced action. So 'scepticism' in that sense is:

> . . . un estado de mente, en que se suspende el ascenso, y ni se determina, ni se desprecia opinión alguna.[14]
>
> (. . . a state of mind in which assent is suspended, and in which no opinion is either approved or despised.)

It is a state of openmindedness, he insists, which requires doctrine to be subjected to examination and experiment, and which allows the scientist not only to agree or disagree with findings of the ancients, but to distinguish between likelihood and possibility, practical proof and likely theory, in the findings of the moderns:

> . . . que la verisimilitud es medio entre el saber y el ignorar, porque la probabilidad, y la noticia no son absoluta ignorancia, pero ni llegan a ser ciencia. Confiesan los Scépticos que tienen alguna clara idea de sus artes, con que no ignoran lo que los demás saben; pero saben lo que los demás ignoran: muy al contrario de los arrogantes Dogmáticos, que afectando saber lo que ignoran, nos hacen presumir que ignoran lo que saben.[15]
>
> (. . . that verisimilitude is a medium between knowledge and ignorance, because probability and information, though not indicating an absolute lack of knowledge, do not denote scientific fact. The Sceptics confess that they have some clear idea of their arts, in that they are not ignorant of what other people know, but they know what others do not know: which is a very different state of mind from that of the arrogant Dogmatists who, affecting to know what they do not know, make us suppose that they are ignorant of what in fact they do know.)

It is a state of openmindedness, he later elaborates, not even provided by Cartesian *Méthode*. For Descartes, when reducing his knowledge to the extreme point of consciousness that the only proof of his existing was the fact of his thinking, was still failing to demonstrate of what, in essence, thinking itself consisted.[16] Even had Martínez been addressing himself to scientists alone he would still have been hampered by the lack of a strictly clinical language commonly used nowadays. Understandably, like his brother *ilustrados*, he was obliged to argue with the Scholastics, as indeed did his model Francis Bacon, in philosophical logic they could comprehend.

Here too, however, certain names and concepts used out of their normal context were themselves inflammatory. Foreign 'heretics' like Bacon or Boyle, normally referred to censoriously, would seem, to all but the most unprejudiced scholars, strange authorities for orthodox Catholics to prefer, just as Catholic authorities had often provoked the distrust of Protestants.[17] While the idea of sceptical 'unteaching' would be likely to offend the prestige of famous University Schools in Spain, how much more offensive, not to say unnerving, that idea would seem when removed from its reasoned context and paraded by polemists before reading men of limited mental means. That this was a genuinely anguished problem to many religious men can be seen in the studies of authoritative scholars outside Spain. For instance, Friedrich Hoffmann who, in his *Fundamenta Medicinæ*[18] of 1695, when feeling obliged, in all scientific honesty, to accept such modern findings as the circulation of the blood, was at pains, in earnest Christian loyalty, to adhere as far as possible to Galenic principles.

Hermann Boerhaave, whose *Institutiones Medicinæ*, which Martínez would also know, was published in 1708, was still in certain respects dependent on Galen's teaching, especially over the remedy of bleeding, and, like Hoffmann, combined much of the new with some of the old.[19]

As regards discussion of details of medical practice, the argument liable to scandalize established Schools was that abstract Logic, by which medical theories had been formulated, was irrelevant from the standpoint of experimental evidence. Possibly a modern approximation to that academic acceptance of formulated theorizing in abstract Logic would be the once blind acceptance of ultra-Freudian theories concerning the physical nature of dreams and womb-memories, of which theories the most highly codified have become outmoded in newer Psychiatry. Another cause of alarm was Martínez's refutation of the firmly established tenet of Aristotelianism that practical evidence was dependent on observation by means of sense impressions: the only means of obtaining practical proof before the invention of modern instruments. Suspicion and distrust of the practical value of such instruments as the microscope, which demonstrated life in a form incomprehensible to normally trusted eyesight, were consequences only to be expected. Martínez in the dialectic character of Hippocrates explains man's confusion in his sense impressions by instancing the fact that a patient can feel the live presence of his leg after it has been amputated. He sums up the matter thus:

> Nuestros ojos son unos microscopios naturales, y no estando en todos igualmente graduados . . . se infiere que apenas habrá dos hombres que perciban los objetos con una misma absoluta magnitud.[20]
>
> (Our eyes are like natural microscopes, and, since they are not equally graduated in all persons . . . one may infer that there will be hardly two men who perceive objects in exactly the same magnitude.)

Among other notoriously provocative features in the general medical technicalities of the *Medicina scéptica* is an attempt to break down the convenient classification of man's physical constitution into four 'humours': (choler, phlegm, melancholy and bile) derived from the argument that if there are four seasons, four ages of man, four elements (earth, air, fire and water) correspondingly there must be four natural humours or conditions in man. Another is Martínez's objection, in the light of Harvey's discovery of the heart machine and circulation of the blood, that the old blood stagnates and naturally or artificially has to be eliminated, and that the source of new blood-supply is the liver.[21] Therefore his call is for a revaluation of the traditional remedy of bleeding. Other Conversations covered by

'Hippocrates' and his companions concern the characterization of disease, the function of Pathology and Surgery, the make-up of the human body, conceptions of treatment, the *rôle* of nature in Medicine, and related subjects under fresh consideration by rationalists.

But outside purely medical matters Martínez, in at least two instances, rightly anticipated criticism of what might be considered his disregard for professional etiquette. Following the precedent set by Boyle, Descartes, and others, he departed from the academic principle of writing in Latin which, not unreasonably, in days of widespread technical ignorance, meant that the general public had been protected from disturbing ideas which it had no means of understanding and so combating. Piquer, in 1752, for example, proclaimed his defence of Experimental Medicine in Latin and thereby escaped some of the notoriety experienced by Martínez. The latter's defensive reason for choosing Spanish, inspired by works of those whom Feijoo called 'los profesores de espíritu libre' (openminded professors),[22] and explained in the Preface of the first volume of *Medicina scéptica*, was that the revered authorities of antiquity, Greeks, Romans, Arabs, had written in their own languages; that one's national language in the circumstances is the least likely in detail to be misunderstood; and that, in any case, Medicine must move into the full light of day to face rational criticism. In other words he was looking for support outside conservative academic Schools. The second breach of etiquette was more serious, for it involved the moral stigma of unseemliness. His open discussion, again following foreign precedent, of manual operations of surgery was as repugnant to academic dignitaries as it was to the self-conscious protectors of public delicacy. The surgeon or practical anatomist had been a hack-technician, beneath academic status, as it were a laboratory demonstrator, whose opinions were not asked, who did not investigate but illustrated certain anatomical characteristics in the interests of preconceived theories, and whose elevation to professional research would be accomplished only by a slow process outside University Faculties.

While scholars may enjoy or at least appreciate discussion of intellectual doubt, the questioning of established standards, supposed facts and realities, revered symbolism and nomenclature, that exercise is likely to cause in the general public, even now, a state of nervous unease. Especially was this true at a time when disturbing ideas, locked less securely than before in Inquisitional headquarters, were being released into the developing Press – the *Gaceta de Madrid*, for example – by announcements of new books; when, after the Bourbon accession, a marked development of external travel broadened academic horizons; and, not least, when the influence of enlightened foreign Catholics, like the Jesuits of Trévoux, trickled over the

Pyrenees. The mind of the *vulgo* was less receptive. The foreign heretic had been rendered suspect on all subjects, for Theology had been the controlling force of all knowledge. How then could he have been entrusted with secrets of the universe, how could his findings be read and adopted without risk of religious contamination? The mere approving mention of Bacon and his sceptically inductive technique was enough to produce on the general reading public, and the 'herd' of semi-scholars, a psychological effect of near shock. So that many of the immediate reactions to Martínez's medical scepticism were genuine expressions of natural confusion.

Not, of course, that the intelligent Spaniard, or any other intelligent eighteenth-century citizen considered that current medical practice was above criticism. For centuries physicians and apothecaries in Spain and elsewhere had been exposed to public ridicule in various forms of entertainment, especially drama and prose satire. For instance, Fielding's satire of surgeons' fundamental ignorance in *Tom Jones*,[23] or Molière's *Le Médecin Volant* (I, ii) or *Le Medécin Malgré Lui* (II, vi) or innumerable Spanish examples of high caricature: Lope's *El acero de Madrid* (I, ix); Tirso's *El amor médico* (I, i); Moreto's *Antíoco y Seleuco* (II), etc, are typical impressions of the intelligent man's view of Medicine before the nineteenth century.[24] Quevedo's assumption throughout the gruesome *Sueños* that physicians, surgeons and apothecaries are more renowned as killers than as curers, is designed to appeal to a public amusement which is relevant to public habits of criticism.[25] Moreover Spaniards by nature had been given to ridiculing, in public or private, any familiar and local institution – outside religion. But there is a difference between common complaint, criticism, or satire against shortcomings in diagnosis or inadequacies of treatment, and an academic challenge to academic Schools of thought. A people does not necessarily distrust the nature of inherited social disciplines because incidental applications of those disciplines are inefficient. Physicians might be thought to be using their lancets or leeches too much, or too little, or at the wrong seasons of the moon, or at the wrong stage of the patient's illness. But that did not mean to the popular mind that bleeding, properly used, was the wrong method of easing inflammation and that it should be discontinued. Physicians in modern practice sometimes provide a placebo as psychological medicine, for the general public still needs the security of recognized forms of pain-killing, sedation or other self-evident means of relief. What the eighteenth-century public emotionally required must have been the security of a medical state of dogmatic reassurance, and what the sceptics were necessarily offering was a school of distrust. Emotional consolations supplied by bleeding, purging, and the taking of supposedly well tested medicines for common diseases, were as necessary to offset physical and nervous tensions of the century as is the taking of aspirin for a twentieth-century headache. Such remedies, like many dangerous twentieth-century drugs, had not been seen to be so totally dangerous that the public could happily countenance their withdrawal.

Moral consciences today may be genuinely disturbed by modern pronouncements on abortion, euthanasia, vivisection, by activity on the frontier between medicine and morals. In the eighteenth century a parallel unease produced by sceptical infection was the result of attempts to divorce medical research from religious and philosophical preconceptions about the nature of matter and the universe. It is true that the average eighteenth-century man of education was at a relative disadvantage in that he had not studied practical sciences at his grammar school. But he is not to be despised as intellectually inferior to his twentieth-century equivalent, nor assumed wilfully to have opposed progress and chosen deliberately to wallow in his obscurantism. Given the state of knowledge of the period he was no less intelligent in his unease than is the man today who worries about the development of atomic and nuclear power, or the transplantation of human brains. Fundamentally the problem then, as now, was concerned with the rights, or wrongs, of scientifically materialist philosophers to extend research to infinity. And the social atmosphere produced by the contemplation of limitedlessness can be as mystically or intellectually stimulating to the few as it is disquieting to the many. Response to Martínez's endorsement of *Epoche* was impregnated with contemporary atmospherics. Accordingly, his own reasoning and that of his supporters and opponents must be interpreted in this context of social nervousness.

A much publicized reply to the letter, as opposed to the spirit, of the *Medicina scéptica* appeared in 1725. It was written by Bernardo López de Araujo, described as a 'médico de los Reales Hospitales General y Pasión' (Doctor of the Royal Hospitals General and Pasión), and is entitled ominously *Centinela médico-aristotélica contra scépticos en la cual se declara ser más segura y firme la doctrina que se enseña en las universidades españolas, y los graves inconvenientes que se siguen de la Secta Scéptica, o Pyrrhónica* (Medical Aristotelian Sentinel against Scepticism, proclaiming the doctrine taught in Spanish Universities as the more certain and stable, and the serious problems instigated by the Sceptical or Pyrrhonical Sect).[26] Here, then, was posted a Sentinel to safeguard general health against wrongminded reformers of the medical standards and 'official' University curricula which Martínez had criticized when referring disparagingly to 'vuestras universidades' (your universities) instead of 'nuestras' (ours). From this possessive adjective of dissociation, together with Martínez's use of Spanish rather than Latin, Araujo pretended to surmise that Martínez was not a University graduate and could not appreciate the need to prevent the public from becoming confused over technicalities beyond its comprehension. This last point certainly was one worthy of consideration.[27] Martínez himself relied rather on the unofficial thinking of the Universities of Valencia, and other outskirts, and on the Regia Sociedad Médico-Chímica de Sevilla (Royal Medical and Chemical Society of Sevilla), mentioned by Peset as preparing the academic ground for modern experimentalism. In his Preface to Volume I Martínez had stated that he wrote not for those within the Schools, where they were in honour bound to defend the authority of Avicenna, but for the 'recién-salidos de la Universidad' (those newly graduated). For Araujo this matter was supremely important. To conceive of medical practice separated from a theoretic Philosophy of Medicine seemed a transgression against age-old laws of civilization. And the more the innovators attempted to explain themselves in philosophical terms – as did Martínez in his Dialogues – the greater difficulty they were likely to experience from a majority of critics able to argue over Prefatory niceties and unable, through lack of technical knowledge, to argue over details of scientific evidence.

Martínez's rather unfortunately phrased self-description as a sceptic could have only one meaning for the average intellectual: the meaning attached to it in classical Philosophy. So that Araujo, with the approval of his sponsors, while generously assuming ignorance rather than heresy in Martínez's argument, devoted most of his defensive protest to proving, so he thought, that the man who believes in nothing cannot be a constructive scientist or a good Christian. To start from fundamentals, he argued, is not a godparent asked, on behalf of the child at baptism, what that child is seeking?, and does not the godparent reply 'Fe es la que busco'(Faith's what I seek)? Is not Christian doctrine based on the *Credo*? Must one not, therefore, first believe, and later reason within that belief?[28]

His objections are echoed by his sponsors. How, one of them asks, can Martínez, a sceptic, understand anything if he cannot define anything? – that is, define anything according to a generally accepted logic. Words, says the same sponsor, such as 'parece', 'quizás', 'verosímil' (it seems, perhaps, probable), in circulation among sceptics including Martínez, can only be described as wishy-washy ('mansedumbre').[29] For Araujo, following the dialectic example unwisely set by Martínez's Hippocrates, his Chemist, and his Galen, argues terminologically to the philosophical end of proving that Martínez is contradicting himself:

> . . . pregunto ahora la definición, o explicación (ellos nada definen) de la Epoche, es a saber: *Es un estado de mente, en que se suspende el asenso; y ni se determina, ni se desprecia opinión alguna*; ¿es verdadera o falsa? Si dice que es verdadera: luego hay cosa cierta, y verdadera, y consiguientemente da asenso a ella, porque no se puede tener una cosa por verdadera sin darla asenso. Si dice que es falsa, ¿para qué la trae? Luego, de todos modos es convencido el Doctor Martínez en sus propias palabras.[30]

> (. . . I now ask, regarding the definition, or explanation [they define nothing] of the *Epoche*, namely: *It is a state of mind in which assent is suspended and in which no opinion is either accepted or rejected*, is that true or false? If he says it is true, then there is something certain and true, and consequently assent is given to it, because one cannot accept something as true without giving assent to it. If he says it is false, then why does he bring it in? So in every respect Dr Martínez is convicted through his own words.)

How, he goes on then, how can Martínez say that there are 'algunas verdades' (*i.e.* físicas) (some truths [i.e. physical]) but that God has hidden an intimate knowledge of them from us? If they are not known, how can they be truths? '. . . ciencia sin principios no hay' (. . . there is no such thing as Science without principles).[31] By which he means that there can be no science in the absence of *a priori* principles.

More cogently, another sponsor summarizes the attitude of the Araujans by saying that the sceptics:

> . . . siendo imposible encontrar la verdad por el áspero camino de la ciencia, ellos alcanzaron a lo menos esta verdad por el fácil estudio de la ignorancia. No alcanzaron las verdades por ser incomprensibles; mas se preciaban de haber sabido y conocido la incomprensibilidad universal de las verdades.[32]

> (. . . it being impossible to find truth by way of the rough road of Science, they reached at least this truth by way of the facile study of ignorance. They did not reach truths which were incomprehensible to them, but they prided themselves on having known and recognized the universal incomprehensibility of truths.)

There was no practical point in Martínez's trying to explain that in using the word 'sceptic' he was referring, not to the fanatical followers of the arch-cynic Pyrrhon, but to 'prudent' sceptics. If Pyrrhon was imprudent, argues Araujo, what evidence is there that other sceptics were prudent? The debater was not, in fact, open to be convinced. But Martínez to some extent had innocently supplied terminological fuel for dialectic fires. It was natural that in the average medical School, and therefore to the average medical graduate, a new philosophy of Medicine and Science involving untried techniques should seem as unsafe, from every point of view, as modern educational methods would seem, or have seemed, to any strict adherent of the three Rs, of Victorian training by rote, of nineteenth-century, Public School obsession with Classical grammar and texts as primary and not incidental disciplines. Also, from the standpoint of established Schools, it was natural to assume that any opposition to accepted patterns and standards must necessarily take the form of what we should now call a militant movement. The eighteenth-century word was 'sect'. To the minds of the Sentinel and his sponsors, Martínez was trying to launch, in their words, a 'secta scéptica', that is, an active organized force of wholesale disbelief, a materialistic School of philosophy which would need to be refuted dialectically point by point. The conception of *Epoche* as a mental state of uncommittedness to any philosophy was too difficult for any organized School to grasp and could become effective only by gradual insinuation.

Meanwhile Araujo proposed to examine Martínez's argument line by line. For that academically derived, textual form of commentary was used by nearly all scholars of the period, including Dr Johnson, whose wit is based on his Scholastic technique of tripping up his opponents over verbal niceties. Araujo cannot be blamed for his inherited method. But from the outset it tied this not unintelligent man to the strict letter of Martínez's words which Martínez himself had confusingly placed in the mouth of his confusingly modernized Hippocrates, with the intention of propagating Bacon's spirit of free intellectual enterprise. A line-by-line commentary, insufficiently related to the author's general context, and suffering from a weak sense of proportion, can lead to anomalies and diffuse misunderstanding. Yet terminological argument, even nowadays, in modern debates, academic or political, is often misleadingly impressive, and a critic's immediate evidence, obtained from illustrations of dissociated details, is easier to memorize, absorb, or recall, than the comprehensive message of the author criticized. In fact Araujo's line-by-line commentary of the *Medicina scéptica* amounts to a line-by-line commentary on merely the Introduction to that work, as Feijoo was swift to indicate.[33] The body of Martínez's text, raising technical questions as to what could legitimately be understood as 'humours' – for example in the Section 'Impúgnanse sólidamente los pensamientos que acerca de los humores defienden las Escuelas' (Firmly opposing the ideas concerning the humours defended by the Schools) – or discussing pathological and surgical details, is disregarded.[34] Where Araujo felt most able to operate was on the once secure campus of Medical Philosophy.

Terminology leads Araujo rapidly to what he and so many regarded as an alarming threat to religion. Martínez himself, in his Catholic circumstances, had almost inevitably taken the precaution of supporting the views he borrowed from such Protestant sources as Bacon by quoting broadminded remarks of the Fathers, especially remarks seeming to belittle the Scholastic authority of Aristotle. In retaliation, Araujo concentrated on all possible dangers attending the word 'sceptic' and its new associates. It had been used in the service of anti-Aristotelianism. But Luther had professed to be an anti-Aristotelian, argued Araujo. Therefore to side with opponents of Aristotelian philosophy as approved by Aquinas was to take sides with Luther. Did Martínez want to be counted among those who follow the arch-heretic? This dialectic bandying of religious authorities in support of religious principles accounts for many of Araujo's poorest arguments, just as it had produced defensively some of the poorest arguments in the *Medicina scéptica*. Don Martín himself was not averse to quoting scriptures, biblical or hagiographical, for his own ends, or employing syllogism when it suited him.[35] Only a creative writer such as Feijoo could free himself from restrictive terminology and Scholastic methods of illustration.

In his Preface, Araujo had spoken urgently of the danger to settled faith of sceptical attitudes, and the risk of the spread of heresy. A sceptic, according to him, operates against Catholic principles; a sceptic is like the man who made no use of his one talent; a sceptic flees from professional Dispute on facts of nature; a sceptic doubts his master and, however innocently, leads man to doubt his God. It might seem all very well for Martínez to protest that his scientific scepticism had no bearing on Theology and the Catholic Faith. The trouble was, contended Araujo, here with undeniable justification, that scepticism in one sphere leads to scepticism in others. The history of French materialism during the rest of the eighteenth century would certainly exemplify his argument. And had Feijoo lived longer he would have been faced with a more difficult task than any he had so far experienced: that of deciding how far it was morally safe to leave *Epoche* to its own devices.

Araujo's sponsors echo his alarm with all the more vehemence because Martínez had provided himself with the support of some items from that authority of authorities, Thomas Aquinas. Referring to this fact, one of the sponsors, Navarro y Aguilar, asserts that modern sceptics, in their ignorance, have merely proved the already established fact that man can only know that he knows nothing, and that even in this respect the sceptics need the help of the very Dogmatists they despise:

> En conclusión los Dogmáticos son los que usan de aquel único modo de saber que ha encontrado la razón: con que buscarle por otro rumbo, o va fuera de razón, o no es modo de saber.[36]
>
> (Finally the Dogmatists are they who make use of that one means of knowing that reason has found: so that to search for it in any other way either goes beyond reason or is not a means of knowing.)

Doubtless it was for similarly logical reasons of religious principle, and not purely out of obscurantist perverseness, that Araujo pointed to the Brief of Benedict XIII, of 1724. This, in theological language, urged constancy in following the doctrine of saints, especially St Thomas Aquinas, against such heresies as justification by faith alone: an injunction which Araujo interprets, apparently in all sincerity, as a warning against any disagreement with Aquinas on any subject. If the Pope could be interpreted as declaring Aquinas' pronouncements in general to be without error, then Martínez, who suggested that scientists were wasting time in studying the *Sumulas*, was, unconsciously, to do him justice, says Araujo, offending against the Papal Brief. However timewasting such arguments may seem to modern readers, we must remember to put them into the context of eighteenth-century Europe and appreciate their serious importance in the whole association of ideas.

The importance to the average intellectual, who had been trained by processes of theoretic definition, of such problems of terminology cannot be overstressed. Free use by scientists of the word 'nature' implied materialism.[37] Discussion of the evidence, or unreliability, or circumstancial abnormalities, of man's senses almost inevitably had to take into account Aquinas' explanation of mysteries of the sacramental species. Since Luther had ranged himself against Aristotle, anti-Aristotelianism, in terminology, as in ideas, could become a threat to dogmatic principles as accepted by the Church. Nor were Catholic thinkers alone in perceiving dangers of changing word-usage. Araujo's argument that philosophical permissiveness can lead to Lutheranism and its doctrine of *gratia per se*, or justification by faith exclusive of works, is one which not only troubled the Anglican Church in general throughout the eighteenth century but produced divisions in the Wesleyan Movement itself.[38]

A related view held by the average European that man was not intended to dabble even in physical mysteries beyond his general comprehension of Christ's words 'I am the way', or to try to penetrate independently all ultimate causes, was largely the explanation of the Dogmatists' belief that approved scientific theory must precede scientific practice, that scientific knowledge cannot derive from the particular but must be directed by God-given laws of Thomasine dialectics for clarification and orthodoxy. Araujo's immediate medical authority here, as in many instances, was the 'divine Valles', 'divine' because Thomasine, and because apparently convincing to theoretic reason.

The statement 'Dios entregó a los hombres el mundo para la disputa, *mundum tradidit disputationi eorum*, que dice el Texto Sagrado, diciendo Valles: *mundum in disputationem* y vea el Dr Martínez como ha de salir de este estrecho' (God gave men the world for disputation, *mundum tradidit disputationi eorum* which Sacred Text in Valles' words says *mundum in disputationem*, world in disputation, and now let Dr Martínez see how he can emerge from this narrow confine),[39] offers an example of how far scientists, according to Araujo, ought to go. For experimental scientists to distrust appearances which have been logically accounted for by reason, Araujo argues, is to propagate nonsense. To suspend judgment in all things, as sceptics have always professed to do, is demonstrably wrong. In his enthusiasm for the 'truth', the Sentinel was talking hysterically beside the point, and outside Martínez's context. Take opium, for example, he says. It causes sleepiness and one may suspend judgment about the reasons for such an effect. But what one cannot dispute is the fact of nature that opium demonstrably induces sleep. He thought he had proved his point that it is dangerous to doubt the outward evidence so carefully observed and interpreted by Aristotle. But the whole of his argument is a dispute about words.

Most of his practical examples are similarly designed to offset Martínez's individual items of illustration, as opposed to the general message of his thesis: his examples, for instance, of misleading outward appearances. Martínez had argued that different people might see the same thing in different ways, be at variance in their conception of colour, etc; that men when mad, or dizzy, see the world in a different focus from that presented to sane men of normal physical balance. They are not perhaps the most convincing of illustrations. For Araujo would argue in reply that, if evidence is to be given of misleading outward appearances of reality, it should not be sought in notoriously false impressions of reality which do not delude a sane and healthy observer. But Martínez was obliged to resort to obvious, extreme examples and could not resort to unfamiliar, specialized evidence of scientific instruments incomprehensible to the *vulgo*. Nor, even on general principles, could Martínez always explain himself satisfactorily. To Dogmatists certain abstract principles such as 'The whole is greater than the part', or 'Cause precedes effect', or the laws of Mathematics which indicate that $2+2 = 4$, seemed so patently inviolable that it might have been wiser of Martínez and his colleagues in Enlightenment to avoid esoteric argument over interpretations of these principles. Like nearly all eighteenth-century scholars, however, he had himself been so thoroughly trained in Dialectics that he deemed it necessary, or found it natural, to fight Dogmatism on Dogmatic ground. So he asserts, when writing against Aristotle's theory of a universal *continuo* consisting of infinite parts, that some part of an infinite could be as great as some individual whole; and that, accordingly, the theory of 'The whole is greater than the part' is incorrect: an argument which, rather understandably, Araujo found 'laughable'.[40] Indeed Hippocrates/Martínez could split hairs with the best. Dialectics properly understood was, said the Sentinel, with an unusual burst of imagination, the key to the door of the house of knowledge which Martínez had tried to enter by crashing through the roof.[41] Again, ironically, Araujo was nearer the truth than he realized. Feijoo, the most dynamic of those who publicized Baconian *Epoche*, was to develop the technique of artistic and intellectual crash and shock to perfection. It was Feijoo, with his keener sense of proportion regarding generalities and particulars than either Martínez or Araujo, who realized, and said over and over again, in other words, that Dogmatists and 'sceptics' were arguing in different dimensions. 'That is not the point'[42] could be said almost invariably of Araujo's objections, and occasionally of those of Martínez.

A typical example of misunderstanding caused by interpreting one medical dimension through the characteristics of another occurs when the Sentinel elaborates on the principles of the 'divine' Valles, his immediate authority. Since God, says Araujo, has given men the true means that is, Thomastic logic, of assessing the ultimate cause of all that is right and possible for them to know in their circumstances, how is it that modern sceptics eschew this only means of investigation? There stood the Araujos on one plane and the Martínezs on another, absurdly accusing each other of failing to recognize the one means of enlightenment. The word *ignorance* or *obscurantism* to each had a different connotation. Dogmatists applied it to lack of training in systematic reasoning. The Sceptics thought of it as absence of objectively tested evidence. And while the two parties continued to speak in different languages, from different standpoints, neither the authorities inspiring Sceptics nor those supporting Dogmatists could be seen in true perspective, still less could Experimental Science be regarded as an acceptable means of investigating truth. From the standpoint of the Dogmatists, practical experiment might be seen to be incidentally useful in so far as it illustrated theory. But so long as it could not be fitted into a new or existing system of universal cause and effect, capable of convincing by theoretic logic, it could not be taken as a serious means of overthrowing existing theories. For this reason experiment, if not regarded as dangerous, though many intelligent men believed it to be so,[43] might at best be considered a secondary exercise of incidental importance. To the twentieth century a secondary exercise, of incidental importance, might seem to be that of theorizing about possible interpretations of evidence. The two *rôles* of Science have been reversed.

The relatively new evidence of the microscope was a case in point. It is with astonishment that the modern reader receives Araujo's information that the microscope distorts the truth, since it makes objects, he argues, bigger than they are in reality. Martínez, discussing the inaccuracy of man's sense impressions, had noted that jasper, which seems smooth to the touch, showed itself under the microscope to have a rough surface. And therefore according to Martínez's argument, pursues Araujo, if the microscope really represented reality there would be nothing smooth in the world: an idea impossible for him to accept. Distrust of scientific technique felt by the average intellectual of those times must have been something akin to modern distrust of scientifically conducted Gallup polls, offering usefully suggestive, incidental pointers which, it is normally agreed, cannot be taken too seriously. Eighteenth-century critics could say with one of Araujo's sponsors, for example, that by no external means could the incontrovertible truth of Transubstantiation be explained. For such considered reasons, and because laboratory experiment was under-developed and much apparatus still unrefined and doubtless often inaccurate, for such reasons, and not normally out of blind stupidity, the Araujos could dismiss experimentation as fallacious:

> El Dr Martínez no quiere acabar de conocer que las ciencias no se fundan en experimentos, porque éstos son falaces de catorce modos, sino en unos principios generales conocidos por la luz de la razón.[44]
>
> (Dr Martínez does not want to come to acknowledge that the Sciences are not founded on experiments, for these are in many ways fallacious, but on general principles perceived through the light of reason.)

To the human reason in general, truth seemed more likely to reside in Aristotle's seemingly reasonable principle of qualities for defining form and matter. Nobody could gainsay his explanation, in itself, of a quality as 'un accidente por el cual somos cuales' (an accident whereby through such means we are such as we are), or therefore object to a description of the eye's 'facultad visiva' (visual faculty) as a fact of universal truth.[45] When the minority of scientists declared themselves to be dissatisfied with such incontrovertible definitions of quality and asked for suspension of judgment over what seemed evident, the majority of intellectuals logically thought them eccentric or perverse.

At its poor best the *Centinela* is an unconscious plea for the means of preserving religious and moral security of mind in the face of a not ill-founded possibility of ethical and religious disturbance and disruption. At its worst, of course, it was an expression of professional prejudice directed against deviations from professional traditions and etiquette.

When Martínez prepared Volume II of his *Medicina scéptica*,[46] it was largely with Araujo and the medical dogmatists, guarded by the Sentinel, in mind. By now Martínez had a more realistic idea of how modern scientific technique would need to be presented in public. Naturally this new series of Conversations discusses some of the Sentinel's specific judgments about man's pursuits of truth, or what he takes to be truth. Conversations turn on the failure of Araujo's philosophical technique to serve for more than theological purposes (C. 29). A refutation aggressively entitled 'Impúgnase la impugnación del Centinela a los errores de los sentidos en particular' (Refuting the Sentinel's refutation concerning the errors of the senses in particular) (C. 30) takes the line of direct attack. The twenty-seventh Conversation answers Araujo's objection to the irreligious nature of scepticism by arguing that the Bible and other holy books favour 'scepticism', in the scientific sense, by implication, that is, by the assumption that Christian 'sceptics' should be ready to distrust their worldly reasoning on which heretics depend. But behind particular cuts and thrusts of particular disagreements there is now a recognition by Martínez of the three general and major difficulties besetting the publicizers of Enlightenment: difficulties more imaginatively and trenchantly voiced by Martínez's wellwisher, Feijoo. They are the linguistic conservatism of popular Catholic expression; unscientific treatment of scientific fact by the recourse to irrelevant theory, authority or material evidence; and the old intellectual habits – by no means remaining exclusive to Spain – of inexactitude, reliance on opinion, hearsay or second-hand evidence, and, most significantly of all, on techniques of syllogistic disputation. That the Defence of the Faith had rested on syllogistic reasoning was a fact which Martínez devoutly underlined. Feijoo, forcefully, and later, Isla sarcastically, tried to show that syllogism and logic, by Scholastic definition, could prove absurdities; could, for practical purposes confuse the mind,[47] that it was not suited to the needs of Experimental Physics, or to matters of Sunday-to-Sunday preaching. As Isla protested to inveterate logicians:

> Esto es, por paridad, como si un maestro de obra prima (que así se llama, no se sabe por qué, a los zapateros), con un aprendiz que quisiese instruirse en el oficio, gastase un mes en enseñarle si la facultad zapateril era arte o ciencia; y si arte, si era mecánico o liberal. Otro en instruirle si era lo mismo saber cortar que saber coser, saber coser que saber desvirar, o si para cada una de estas operaciones era menester un hábito o instrucción científica que las dirigiese.[48]

> (By analogy this is as if a master of the prime trade [for so shoemakers are called for some unknown reason], when teaching an apprentice wanting to learn how to do his work, should spend a month teaching him whether or not the shoemaking faculty was art or science, and, if art, whether it were mechanical or liberal. And [as if he should spend] another month in teaching him about whether knowing how to cut was the same as knowing how to trim, or if for each of these operations would be needed a customary method or scientific instruction to direct them.)

Yet reasonably, for the period, syllogistic logic, as an accepted discipline of thought, seemed as dangerous to set aside as it would now seem dangerous for any country to set aside the laws of its Constitution, or a university to change its standards of values, or for a modern political party, in critical parliamentary debate, to concede points to opposing parties by foregoing terminological needling. Such habits of mind cannot be dismissed by edict, argument, or even always at will. The 'paroxysm of laughter' provoked in him, so Martínez said in retaliation to some of Araujo's pronouncements,[49] was unlikely in the circumstances to further Martínez's cause. It is difficult to submit to derision. But Martínez's paroxysms were mild and infrequent by comparison with those of Feijoo, and he was more disposed to reason with his intellectual inferiors than to shock them.[50] Not for him were the alarming paradoxical statements used by the Father Master to shake his readers into attention, or the blackmailing threats of, for example, the Astrologer Torres Villarroel who, in response to Martínez's strictures against astrological prognostications, threatened, if Martínez continued, to publish statistics of the Doctor's patients who had failed to survive their illnesses.[51] Consequently, had it not been for Feijoo's intervention, Martínez's 'scepticism' might have remained only a nine-days' wonder.

Without exactly determining the course of polemic concerning scientific Enlightenment, the *Centinela* and the Conversations of the second edition of *Medicina scéptica* helped to influence the way in which future polemists would travel, and to a certain extent they foretold the lines and projected the tone of the specialist *versus* non-specialist debate. Feijoo, who as a literary genius in his own right, owed nothing, stylistically, to anybody, was nevertheless obliged by the climate of circumstantial unease, to move polemically in similar directions to those followed by Martínez. The same theological scrupulosity in his opponents, their same methods of defence, or personal attack, their captiousness out of context and off the point, above all, their ignorance about modern scientific scholarship, and their conservatism in matter and technique, were experienced by the Benedectine on a large scale proportionate to his larger personality and consequent publicity. The precise nature of his subject matter was also partly determined by polemic in progress. Perhaps he learned some of his modern Medicine, acquired some of his medical bibliography, and chose some of his scientific illustrations from the *Medicina scéptica*, though these belonged to a common fund of international scholarship on which Feijoo drew even more widely than did Martínez. Following Bacon, and doubtless reminded by Martínez, he was fascinated, for example, by new discoveries in Optics,[52] which a section, called 'Falacia de sentidos' (Fallacy of the senses), in Martínez's 'Conversación 29' may have encouraged him to develop. He seems to build on Martínez's references to experiments with a thermometer in a cave, and he provides a more vivid equivalent for Martínez's criticism of Dogmatists who, through lack of experimental training, speak vaguely and unscientifically of 'qualities'. One might as well say, Martínez argues, that mules can draw coaches because they have a 'cualidad tractriz' (a drawing quality).[53] One might as well explain the cause of heat, Feijoo echoes, in a better example, by saying that it has a 'cualidad calefactiva' (a quality of calefaction), or, improving on his own analogies as he is wont to do, by saying that the movements of a watch are caused by the 'forma artificiosa de la máquina la cual tiene virtud artificial para causar esos movimientos' (artful form of the machine which has an artificial capacity for causing these movements).[54]

On religion Martínez was as necessarily obsessive as was the Sentinel. Perhaps Martínez by now had realized belatedly that terminology had driven him on to slippery ground. For the next few decades the significance of the word 'sceptic' would preoccupy critics, enlightened and unenlightened alike, who might have been more fruitfully employed in investigating other forms of knowledge. Self-justification became an overworked necessity whose tedium was only relieved by the ironic laughter of Feijoo in the area of informed scholarship, and by the hilarious malice shouted confusedly, if artistically, by Professor Torres Villarroel in the area of pseudo-scholarship. We tend to subordinate the arguments of these two literary geniuses to their challengingly artistic self-expression. But, in their own day, their raised stylistic tones related to urgent ideas and emotions, and contributed to a rise in the pitch of public unease, thus publicly representing the century's nervous crisis of thought.

For his part Martínez, in self-defence, is careful to spell out to his prosecuting 'Licenciado Vidriera' (glass-frail Licentiate) of a Sentinel the purely secular significance of his scientific scepticism by showing, from Araujo's own arguments, that the latter neither knew the history of the Sceptics nor was prepared to study *Medicina scéptica* or correctly interpret its title. It is as if, says Martínez through the 'amused' Chemist in Conversation 26, it is as if the title read 'Sceptical Theology' or 'Sceptical Metaphysics'. In these circumstances Martínez was right to say that the Sentinel's attack was against not a book, but an author, against a purveyor of unfamiliar concepts, one who, to quote Araujo's School of Authorities, denies the precept that 'el instituto de las Escuelas es defender los dogmas de Hippócrates, Galeno y Avicenna, con tan rigurosa obligación como obediencia deben tener los católicos a la Santa Sede' (the object of the Schools is to defend the dogmas of Hippocrates, Galen and Avicenna with that strict obligation which is comparable to the obedience owed by Catholics to the Holy See).[55] These critics who know nothing, says Martínez, of discoveries in Optics and Anatomy, should remember that St Augustine – the reference to such obvious holiness constituting for the times the strongest of arguments – could have doubts concerning explanations of the physical universe and regard science, unlike religion, as an area of inquiry: an Augustinian attitude coinciding exactly, concludes Martínez in triumph, with a true medical scepticism.[56] These medical critics, to his mind, were they who, secure in their world of theories, did not descend to practical matters affecting diet, the interrelation of diet, drugs, medication and the course of diseases. These were they, Martínez's Chemist had noticed, who could be so ill-read in Science as to confuse Robert Boyle with Francis Bacon, calling the former 'El Gran Canciller de Inglaterra, Roberto Boyle' (The Great Chancellor of England, Robert Boyle).[57] These critics are to be envied, mocked Martínez's Chemist, for their way of proving their case without saying anything substantial, being accustomed to 'assume as evidence what they need for evidence'. These are they, sums up Hippocrates in crushing understatement, who differ from Medical sceptics only in so far as 'saying differs from doing'.[58] These were they who regarded medical logic as a syllogistic battle-ground, solemnly debating such questions as 'What is pain?', when they should have been investigating clinically the physical causes and effects of the phenomenon pain. These were they who searched for those unknown components or functions of nature which, as unknown, are, unlike religious truths, so far undefinable.[59]

Naturally enough Martínez brought up for discussion those new European doubts about the accuracy of man's sensual observation on which the Araujan majority relied. It is now difficult for a reader to appreciate the bafflement which the evidence of the new telescope, the microscope, and the experiments of ophthalmic scientists occasioned in those who were told that their senses could deceive them in ways never before imagined. This was not merely in the way in which nature visibly deceives the senses, for instance by making a straight stick look twisted in water, but by means, invisible to researchers without mechanical aids, of proving the movement of earth round the sun, and the existence of microbes. Readiness to distinguish between an Aristotelian conception of observation by the senses and that of experimental physicists required a process of re-education.

On purely religious matters, however, Martínez's disparaging 'paroxysms' of laughter sound a little hollow. By now he could better assess the force and complex forms of opposition which hitherto his practical mind could hardly have anticipated. The Sentinel set the tone for popular reaction to the reforming Spanish scientists, and Martínez and Feijoo set the tone for intellectual retaliation, that inflammatory word 'sceptic' acting as a warcry on either side. In fact neither pro-sceptic nor anti-sceptic was fighting an enemy whose challenge he fully understood. The scientific expert now was not merely engaged against fellow experts in learned journals. The dogmatists now were not fighting other dogmatists over matters of theoretic principle. The two groups spoke different languages. Inevitably, as the engagement developed, it was characterized on the one hand by sarcasm and outrage and, on the other, by splenetic mockery. In the interests of self-defence, especially on the dangerous ground of religion, the Martínezs, Feijoos, Islas, were obliged to conduct personal attacks as forcefully as was within their power. It was not sufficient for Martínez to protest that he doubted only in spheres in which nature's truths had never been explained, or which had been left open to mere opinion: 'dudo de lo opinable . . . negamos las cosas de especulación, admitimos las verdades prácticas' (I doubt what is opinionable . . . we deny matters of speculation, we admit practical truths).[60] It seemed necessary, wastefully, we should now think, to pass condemnation back to the accuser. So, to use Araujo's own method of argument, could not he, the Sentinel, be called heretical for following blindly the heretical Aristotle who not only belonged to heathen times, but was called sacrilegious by the earliest Christians?[61] Support for Bacon, Sydenham, or any modern experimentalist must needs be strengthened by quotation, as relevant as possible from the Bible, the Saints, particularly Saints Augustine and Thomas Aquinas, from Papal Bulls, or from other theologically sanctified authorities. The method was time consuming and conducive to acrimony. But, despite Bourbon support and the support of influential clergy, it was the only method, until later decades of the eighteenth century, if then, which the 'herd' could understand. Indeed the only reason, explains Martínez, why he has troubled to reply to the ignorant Sentinel at all, is:

> . . .por no otorgar con el silencio la sacrílega nota de casi herética que me imputa . . . comentando mis ideas siniestramente, e intentando persuadir que están en mi mente cual él las dice.[62]
>
> (. . . so as not to authorize by silence the sacrilegious, almost heretical tone which he attributes to me . . . commenting on my ideas perversely, and endeavouring to prove that those ideas which he says are in my mind are in fact there.)

In face of all provocation, Martínez goes on, he will try to reply with moderation to a critic who has only to hear the words 'doubt' and 'sceptic' to rush into print 'with the first thing that comes into his head'[63]: a remark heightened by Feijoo's indictment, delivered with more punch, of the protests of a Capuchin critic in which 'a cada paso se encuentran embolismos, en que no se percibe por donde va, ni para donde viene, ni aun si va, o si viene' (at every step are encountered embolisms whereby it is not clear where he is going or for what reason he has come, or even whether he is going or coming).[64]

Publicity for Martínez's *Medicina scéptica* was more dramatically increased through the lively personality of Feijoo, whose defence of its author, written from Oviedo in 1725, was republished in Martínez's third edition of 1748, again as an 'Aprobacián apologética' (Sponsorial Approbation). By then Feijoo had become a force to be reckoned with in polemic, and his disdainful treatment of López de Araujo raised the temperature of acrimony. Not that Feijoo can be blamed for his disgust at 'pigmies' who 'take on giants'. The 'ineptitude' of Araujo's arguments, the dangerous 'irrelevance', particularly concerning that word 'sceptic', his inappropriate references and quotations, even his 'poor style', 'aunque con inútiles esfuerzos de culto' (despite his futile attempts at sophistication), are natural retaliations to one who, the Father Master mocked, had used 'decorous' adjectives like 'stupid', 'crazy', 'ignorant' about the enlightened Martínez.[65] Feijoo's tone may not have encouraged rational acceptance of his argument. His remark that Martínez's Introduction had been almost enough to make Araujo throw up his bile, was hardly conducive to quiet discussion. But, as usual, he went straight to the roots of his victim's inadequacy and seized on the fact that Araujo had concentrated on Martínez's Introduction to the exclusion of his textual explanations, and had taken fright at words which he, Araujo, senselessly removed from their contexts, beginning inevitably with the word 'sceptic'.

Feijoo was in a better artistic position than most polemists to appreciate the pitiful irony of such irrelevance and to convey it caustically. History cannot supply the Araujos with any logical defence. Yet they illustrate a natural confusion of mind. Emotive words of themselves are liable to generate both self-righteous and altruistic indignation. The average man, insufficiently prepared to enlist in the front ranks of Reformists, had a humanly reasonable fear of the spread of uncontrolled scepticism, and instinctively attempted to restrict, discourage, or at least question the new tendency to make doubt the subject of propaganda and give it independent life. On the other hand, his provoking habit – a common habit, humanly speaking – of reacting to words rather than to ideas, and so of arguing beside the point, accounted for the bitterest and most personal quarrelling of eighteenth-century polemists and would contribute to Fernando VI's decree forbidding Feijoo's critics to retaliate. Had scholars used a less inflammatory word than 'sceptic' in a country as defensively Catholic as Spain, the spilling of much bad ink might have been avoided. But that word, though the most notorious, was not the only watchword of strife. Nor was Feijoo, whose personality thrusts itself more vigorously through his writing than did that of Martínez, disposed to moderate his vocabulary for the sake of peace. Feijoo was a natural fighter, unafraid of dangerous weapons.

It is significant, however, that Feijoo took Araujo's misdirected criticism with sufficient seriousness in the prevailing context of scientific tension to realize the importance of examining that term 'scepticism', and explaining the difference between its ancient and modern usages. Just as surely, he had grasped the importance of detaching Aristotle from the enthroned position in which St Thomas Aquinas' use of his methodology had placed him, and which had been further guaranteed by the Brief of Pope Benedict XIII declaring that St Thomas' works are free from all error. The assumption for religious reasons that Aristotle, like Hippocrates and Galen, must not be contradicted was in Spain peculiarly difficult to eradicate, above all in jealously religious Colleges and University Schools. The idea that Aristotle's Physics and Galen's Medicine could, by modern instruments, be faulted, might be readily acceptable to a minority of scholars, but the average man, then as now, had difficulty in restraining his judgment until he was in possession of all the facts of the case, for example that Aristotle could be proved partly right in his *Dialectics* for theological purposes, and wrong, in his Physics, for scientific purposes, without offence to St Thomas or to the Holy See.

Apparently the negative preparation of mind for major changes, the suspension of judgment, was more difficult of acceptance than was the positive practice of the new Science. It was the new explanation of phenomena rather than the presentation of new phenomena – such as new stars, microbes, electric current – that raised alarm, figured most conspicuously in polemic, and so occupied most urgently the attention of Martínez, Feijoo, and their supporters. The popular favourite Torres Villarroel, to be discussed in more detail later, represents a class of leaders of men who did battle for a modified principle of enlightened Science in the armour of tradition. When Feijoo lost his patience in print, as he often did, it was usually because he was having to contend, not with a prejudiced or muddled individual, but with a prejudiced representative of a whole prejudiced class. In which case, therefore, it was virtually impossible to conduct discussion by the distrusted, unpopular, supposedly negative means of *Epoche*. But, in his defence of Martínez, Feijoo did not fail to emphasize the positive and concrete aspects of enlightened medical thinking. He underlined Martínez's demand for the reform of medical education, of clinical practice, of examination systems, of concrete, physical analysis, and of instruction on scientific experiment rather than training in the defence of abstract theses.[66] Positively, too, as a good teacher, he supplied examples for his argument. How does Martínez's maligned scepticism work? It works, for example, he says, not by his denying recognition of proved phenomena – such as the traditional efficacy of rhubarb for purging – but in puzzling openmindedly about the reason why rhubarb should operate as it does. Scepticism works too in the scientists' caution, by their pondering over the physical causes of certain known realities, such as the circulation of the blood, proved, but as yet inadequately explained. Behind the sceptics' unwillingness to categorize the workings of the physical universe is the knowledge that no school of scientific theorists has been able to prove itself against the rest. For which reason a critic's doubt or fear can be healthier, insists Feijoo, than blind adherence to any one of the various schools of ideas. Still, however much sense Feijoo's Apologia might make to those willing to listen, his tone, individualistically astringent, was in itself a challenge. It assumed a manner of individual thinking that could lead a man out of his known world into mental discomfort and loneliness. The average mind needs ideological security. Stylistically Feijoo reached his climax of defence, not in his positive plea for positive College reform, but on a mind-piercing note of disparagement over the unsceptical confusion of Science and Religion:

> En esto se fundan algunos extranjeros cuando dicen que en España patrocinamos con la religión el idiotismo.[67]
>
> (Some foreigners are basing their reasoning on this when they say that in Spain we make religion the patron of idiotism.)

Neither seventeenth-century England nor eighteenth-century Spain could be changed by quiet persuasion. There are times when shock tactics are essential.

The polemical confusion of cross-purposes provoked by talk of 'scepticism' and all that that word could imply about materialism, foreign heresy, and moral depravity, was exteriorized as a paradox with the realism of nightmare unreality by Torres Villarroel when he entered the polemical arena in the 'twenties and thrived artistically on current strife and exasperation. Torres might be said to transfer to ink, as Goya transferred to the canvasses of the *Caprichos*, the common tensions vibrant in perplexing change. Despite much outward bravado of scientific appearances, contrasting with his instinctive resort to inherited authority and his practices of folk-quackery, Torres never wholly committed himself either to conservatism or to the new rationalism, and so was more representative of the thinking and feeling of his times than he realized or would have wanted to be. Like his fellow spectator, Goya, he intuitively converted the formlessness of the current confusion into concrete symbols of deformity, by means of the Golden Age, the Quevedan technique of splitting the illusion of sanity and linking unlikely parts of it together. This he did, for example, by exteriorizing one sense impression in terms of another ('mugrienta luz', 'Parecióme que traía el alma en remojo', 'hemos . . . hacinado tanta escasez') (greasy light, It seemed to me that the soul was in soak, we have stacked up so much scarcity),[68] by dramatizing the confrontation of opposites, and by concretizing atmospherics or nervous apprehension ('Los huesos se me metían unos dentro de los otros al oírle . . .') (On hearing him my bones got one inside another).[69] It was a technique commonly used by Quevedo, for example, in his description of the strange behaviour of mixed human remains at the Trump of Doom,[70] or as reflected later in Goya's nightmare-like contortions. All of which, therefore, far from being new and pre-Romantic as has sometimes been assumed, is a natural development of the late Renaissance search for the inexplicability of ultimate reality. This might entail a philosophical search, as through the paradoxical implications of *Don Quijote*; or a religious search, by means of San Juan de la Cruz's ascent through 'nothingness' to a sixth dimension; or a materialistic search, as in Quevedo's *Sueños*, or an almost purely emotional search, as re-emerging in the receptive Torres and Goya.

One of the ways in which Torres helped to confuse the ideological issue was by following the fashion of Bourbon enlightenment to call himself sceptical, or 'doubting' ('incrédulo').[71] Muddled thinker, unsound scholar and opportunist as this artist was, he ignored, or was ignorant of, the difficulties involved in accommodating his 'scepticism' to his folk-mentality. By 'doubter' he really meant that he was sceptical of things not obvious to common observation, or of ideas inconvenient to him, such as the irrelevance of Astrology to Medicine. Another satirist of Torres' own artistic calibre, and as vicious a fighter as he, was the Jesuit Father Isla, who had supported Martínez and Feijoo and who played so significant a part in another branch of polemic.[72] In his *Glosas interlineales* Isla jeers at Torres' confusion in his self-denomination as a scientific sceptic, by invoking the presence of that enlightened sceptic Feijoo at Oviedo:

> Pero en todo caso yo me holgaría, que ya que Vmd (*i.e.* Torres) en Prólogos, Papeles y Diálogos se ha declarado scéptico, fuese de los nuestros que se le daría indulgencia plenaria, y remisión de todas sus Postdatas, sólo con que dijera conmigo *Padre nuestro que estás en Oviedo.*[73]

> (But at all events, I should be satisfied, now that in Prefaces, Articles and Dialogues you have declared yourself to be a sceptic, if to become one of our

> kind, you were to be given plenary forgiveness and remission of all your Postcripts simply by saying with me *Our Father who art in Oviedo*.)

The state of social puzzlement or of muddled paranoia can be a vital, sometimes an inspiring mental experience, and an artist who responds to the phenomenon for its own sake, that is, with no clear notion, means, or even conscious or unconscious desire to contribute to its remedy, is perhaps its best interpreter to posterity. Thus, when Professor Torres who, for all his inadequate training, liked to fancy himself as a medical scientist of Enlightenment,[74] broke into the Martínez-Araujo controversy, it was to echo emotional cries against foreign scepticism, and, in an orgy of self-expression, to take artistic advantage of ideological upheaval. His folk-mind, in part a mind dependent on inherited assumptions, in part inclined to emotional opportunism, reflects the rate at which new ideas, through distrust, or confusion, or *guerrilla*-warfare of personalities could gradually become absorbed. Moreover Torres' popularity with the general public, his very talent for satire and caricature, was a force in polemic which often gave importance to his irrelevance, substantiated his emotional arguments conducted off the point, and contributed to an acrimony which delayed the general acceptance of plain fact.

Many polemists had been outraged in all honesty by the terms 'sceptical' or 'unteaching' and by any idea of reliance on 'heretical' scholarship. Torres certainly was not one of their serious number. His personal attacks were shaped by his personal needs of the moment. Feijoo, as an influential Benedictine, was not an easy target. Torres' attitude to the Church was always obsequious. But, in 1726, Feijoo had devastatingly inveighed against the claims of medical astrologers.[75] He had threateningly reminded them that the Bull of Sixtus V condemned astrological prognostications, other than those concerning weather and related matters, as evilly superstitious. Whereupon the astrologer Torres, unable to contend with Popes, had retaliated by criticizing Feijoo's unmonkish activities, for instance:

> El religioso entró en sus claustros a comer mal, y a azotarse bien, a esconderse del siglo . . . Déjese de escribir contra médicos y astrólogos . . . Que el Padre se meta a médico sin licencia de la Bula, es buscar irregularidades . . . Escriba contra las desórdenes de la carne.[76]
>
> (The monk entered his cloister-world to eat badly and to flagellate himself well, to hide away from the world of his times . . . Let him stop writing against medical men and astrologers . . . For this Father to engage in medical matters without permission through the Papal Bull is to engage in irregular activities . . . Let him write against the disorders of the flesh.)

Which, with its implication of religious unorthodoxy and impropriety turned academic debate into *guerrilla*-slander.

Artistically Torres could beguile his readers into agreeing with him. When, in 1752, he declared, on a calculating note of confidence, that in his capacity as a Mathematical teacher, he, supported by his students, had achieved his task of rousing Spain from her sleepy sickness:

> Con estos alientos, y la bulla, y el ruido de mis roncos Kalendarios conseguimos despertar a la España de la modorra en que yacía . . .[77]
>
> (With this spiritedness, and the excitement and noise produced by my hoarse Almanacs we managed to waken Spain from the sluggish state in which she lay.)

he was demonstrating one of the ideological ironies of his times. For his way of rousing Spain from intellectual sleep was by combining the acquisition of accurate globes and books on Geometry and Astronomy with the furthering of astrological prognostication, to the supposedly greater glory of Science. In other words, the 'sceptical' Torres played with terminology for his own ends. To judge by his accounts in the *Vida*, which, in this instance, may well have been accurate, he was popular with his students and an entertaining teacher.[78] Yet an easy popularity derived from easy exposition of difficult academic matters often spells academic superficiality. Certainly Torres' mixture of many old beliefs and opinions with a few modern facts, served up together in a sauce of judiciously 'enlightened' jargon, was a concoction which his mid-century admirers found easier to digest than the medicine prescribed by Feijoo.

But Torres, however conspicuous as a public idol, was not alone in propagating an inconsistent, half-way 'scepticism' of convenience. Another member of the conservative University of Salamanca, Francisco Suárez de Ribera, replied to Feijoo in 1727 in a paradoxically entitled *Escuela médica convincente, triumphante, scéptica, dogmática, hija legítima de la experiencia, y razón* (*The Medical School, convincing, triumphant, dogmatic and legitimate daughter of experience and reason*).[79] Like Torres, this Galenic apologist was anxious to prove his modern scholarship, his scientific acceptance of reason and practical evidence. He was a more serious and courteous, apparently a better intentioned and more fairminded critic than the subjective Torres. His eulogistic sponsor, a member, like Martínez, of the Regia Sociedad Médica y Chímica of Sevilla, who writes the *censura* (Censorial Licence), underlines Ribera's appreciation of Feijoo's general desire to rid medical Faculties of false ideas, and to stimulate openminded discussion and reform:

> El Rmo. Padre sólo ha intentado, desterrando errores, proponer sus dificultades para que levantándose sus profesores del sueño de la ignorancia hallen con su trabajo en la ciencia los logros de la mejor aurora.[80]
>
> (The Most Rev. Father has only tried, by banishing errors, to consider the difficulties [in Medical Faculties] involved for their professors in rising out

of the sleep of ignorance, to find, through their work in Science, the benefits of the finest dawn.)

This was the kind of mental activity, he adds, which disproves Barclay's statement that Science in Spain is unenlightened.[81]

Indeed, in the intention of his generalities Ribera sounds reasonable and doubtless meant to approach scientific Medicine reasonably. He might think that Feijoo had gone too far, farther, he wrongly thought, than Martínez. For Martínez's only protest to Feijoo's strictures against modern Medicine had been concerned with the Benedictine's suggestion that patients would do best to dispense with medical treatment altogether.[82] Otherwise he was as distrusting, as 'sceptical', of medical traditions as Feijoo himself. And Ribera's possibly calculated misinterpretations of the Martínez-Feijoo positions is a typical example of eighteenth-century argument. It was, as Feijoo would say, 'off the point'. But otherwise Ribera's argument, initially and superficially, seems unprejudiced, and his preparatory statements could hardly be faulted by rationalists. They include assertions that it is necessary to be sceptical of dogma, to employ reason and experience; that discussion in modern society need not be ruled out; that a tone of respect and courteous deference in any criticism of Feijoo's 'Discursos' does not imply timidity.[83] Here was an intelligent call, apparently, for quiet consideration and discussion, like Martínez's own respectful consideration of some of Feijoo's suggestions, at a time when any disagreement was assumed to signify condemnation. Thus far, then, Ribera admits the fresh air of objectivity into the close confines of the early eighteenth century where Feijoo's protest 'Why cannot I say I don't know'[84] was normally regarded as inadequacy.

It is when Ribera courteously exemplifies his criticism of Feijoo's 'Discursos' that his idea of scepticism, like that of Torres, is seen to bear no relation to the word as used by Martínez and Feijoo. For this word, even while acquiring with years a certain fashionable acceptance, tended to become fashionably restricted to its user's subjective simplifications. All that Ribera's 'scepticism' amounts to, we eventually discover, is a willingness to concede that Medical Schools and their individual exponents are not invariably infallible, despite the ultimate infallibility of inherited medical authority; that Galen and Hippocrates can be misinterpreted; and that many medical practitioners are inadequate and 'stupid'.[85] This means that Ribera, along with other well intentioned advocates of medical reform, was incapable of regarding the subject of Medicine or Science in other than philosophical terms. One of many of his disastrous interpretations of 'reason' and 'experience', as applied to medical evidence, is his 'proof' that bleeding had been sanctified to medical use by biblical authority. Now, if the Scriptures could be found, in some practical way, to sanction such a practice, the 'scepticism' of the Feijooists would be shown to have reached the limit of its permissible function and be obliged to retreat before rightful dogmatism. Feijoo, often obliged to argue within the confines of his critics' mentality, had protested that there is no reference in the Bible to bleeding and purging.[86] Wherefore, ran his sceptical train of thought, dogmatic critics could claim no biblical backing: a point taken by the not unintelligent Ribera. Nevertheless the latter could show, he asserted – and this is where the 'practical' medical knowledge of 'reason' and 'experience' reverts to philosophical theory – that bleeding had been exemplified in the facts of the Crucifixion. Christ, on the Cross, had given his breast to be pierced by the soldier's sword, and had thereby sanctified the evacuation of blood for the health of the human race:

> Cristo, nuestro Señor, por ser médico tan piadoso, tuvo bien que estando en la Cruz le sangrasen de su pecho sacratísimo, para que . . . luego bien podemos de aquí tomar ejemplar, y hacer tránsito a romper la vena con otra lanza más pequeña, para que haciendo evacuación de sangre, se redima la salud temporánea del cuerpo . . .[87]

> (Christ our Lord, as so pious a medical doctor, thought it well that, while on the Cross, they should bleed His most sacred breast, so that . . . then we may well use this as an example, and go on to pierce the vein with another, smaller lance, whereby, producing an evacuation of blood, the temporary health of the body may be redeemed.)

Confutation of evidence like this could be argued out only on theological grounds, it will be remembered, and these, despite the predisposition of Bourbon statesmen and enlightened Spanish Churchmen to divorce Science from Theology, were, if not precisely dangerous, at least too precipitous for comfort. It must also be remembered, however, that the Riberas of Catholic Spain, average men of the scholar-herd, had their equivalents abroad. These too were well intentioned men of 'reason', as they would style themselves, who, in one way or another, moulded the Scriptures to their own purposes, scientific or otherwise, and found religious interpretation for everyday phenomena in an honest fervour of uncritical faith. Whereas John Woolman, like John Wesley, genuinely believed on so many occasions that rain had been stopped by heaven's decree in order to let him preach effectively in the open, or that God had quelled storms that beset him on his way to missionary destinations, Feijoo, before believing, would have asked for more evidence. The average foreigner was not as objective as the above-average Feijoo or Martínez. From the evidence of the Benedictine's normal method, his response to such statements as those of Wesley and Woolman would have been to record, from their own accounts, how often rain and storm had not conformed to their expectations, and to have suggested that they would be well advised to regard at least some of these apparent miracles as coincidences.[88]

Woolman, like Ribera, interpreted material phenomena by means of non-material and religious assumptions which admit us to inner recesses of eighteenth-century normalcy. Ribera, like Torres, was a 'sceptic' in name rather than in nature. But he typifies a growing number of Spaniards comparable with growing numbers of foreigners who, while unwilling or unable to reform their techniques of thought, were unwilling to consider themselves ill-informed or out-of-date. The logic of their half-way 'scepticism', sincere undoubtedly in Ribera's reasoning, if merely opportunist in that of Torres, is so much more representative of the average mind than is the advanced thinking of internationally renowned individuals, that only on this herd-level can the general atmosphere of the times be understood. Only on this level can the difficulties with which the enlightened had to contend be satisfactorily analysed. For instance, as regards Medicine, the representative Ribera illustrates the naturalness established by both academic and folk tradition of linking disease and the medical remedies with astrological conditions. Yet they both attempt to explain their outdated principles in accordance with changing jargon. Thus Ribera, to offset strictures by Feijoo and Martínez against the obscurantism of relating medical treatment to the position of stars, and of confusing Astrology with the experimental science of Astronomy,[89] modernizes his thought to the extent of re-defining the word 'Astrology'. The science which Medicine must take into account, he says with the satisfaction of one moving in the van of knowledge, is 'astrología astronómica' (Astronomical Astrology), a science necessary 'para guardar con rigor los plenilunios, novilunios etc. en punto de administrar los remedios indicados' (strictly to observe [the times of] full moons, new moons etc. for the timing of administering the remedies indicated).[90] The tendency to understand only half of a new idea, or of newly discovered evidence, and to express it in wrong categories of language, is probably the commonest, the most natural tendency in any period. It is also one more difficult to eradicate or refute than wholesale ignorance and unyielding conservatism. Torres, Ribera and their kind were bolstering their not unreasoning minds against the threat of total uncertainties. In another treatise, *Medicina cortesana*, of 1726, Ribera explains that he is disturbed, not indeed by Feijoo's disgust with 'idiot' medical theorists and practitioners, but by his argument that even 'non-idiots' can be wrong. With syllogistic earnestness Ribera denies such a possibility:

> En los sólo eruditos, concedo: en los eruditos y experimentados con mucho ejercicio, cargados de experiencias muy firmes, niego.[91]

(Regarding the sole erudites, I concede; regarding the experienced, well exercised, and thoroughly and steadily practised erudites, I deny.)

Opposition to Martínez and Feijoo, then, was not necessarily unfruitful. At least it familiarized the scholar-herd, if not the general public, with new ideas which could later be taken for granted, like the controversial word 'scepticism' itself.

Between 1726 and 1730, while echoes of this chilling word still vibrated, a number of lesser known protesters wrote, at length or in passing, about their fears over modern attempts to separate Science and Religion. In 1726, for example, an anonymous 'Barbero' (Barber), a 'Médico de Sarabillo' (Medical Doctor of Sarabillo), and a fancifully named 'Eustaquio Cerbellón de la Vera', object with varying degrees of emphasis that a member of a religious Order – the Benedictine Feijoo, in fact – should sponsor Martínez's doubts and distrust, or meddle at all in non-religious subjects. It must be admitted that, from a polemical standpoint, the exclamations of scandalized astonishment that a clergyman should engage in worldly debate were a convenient means of personal attack used in Catholic and Protestant countries alike. But there is no doubt that the blessing given by Feijoo to Martínez's 'scepticism', however well he qualified that word with such adjectives as 'moderate', was more likely to disturb than to placate. Nor were all of the protesters blind partisans of the Sentinel. The 'Barbero', so called to emphasize his reaction of common sense, objects to the unseemliness of both sides of the polemic in which Martínez and Feijoo, Torres, Pedro Aquenza and Ribera have engaged, and he criticizes quibbling for quibbling's sake. Pedro Aquenza, a royal physician, speaks in alarm of the crisis occasioned by the general scepticism of Feijoo's *Teatro crítico universal* as 'una lluvia de crisis . . . y se infiere de lo llovido en su primer tomo, que podemos temer un diluvio' (a rainfall of crisis . . . and it can be inferred from the amount of the rain fallen in his first volume that we can fear a [future] deluge). All of which is to the benefit, Aquenza believes, of nobody, since – and here he touches the nerve of herd-unease – since Feijoo cannot set up a surer system for the one he is decrying:

> Un público universal Theatro Crítico se ha esparcido estos días a fin de manifestarse algunos tan rancios como tolerados deslices, sin proponerse medios para levantarse los caídos, ni prevenciones para no tropezar en adelante los por caer.[92]

> (A public universal Theatre of Criticism has spread itself abroad recently with the aim of exposing certain weaknesses, antiquated yet tolerated, without proposing means, for those [scholars] who have fallen, to lift themselves up, or without giving warning against stumbling for those who in the future are likely to fall.)

Ribera's underlying sympathies are largely with Torres, the truth of whose prognostics he supports, he says, from observation, and in part they lie with the medical dogmatist Aquenza, whose engagement against Martínez he nevertheless deplores as beneath the former's dignity. But most clearly out of his general condemnation emerges his surprise both that a religious should engage in scientific controversy – 'que un religioso grave . . . se meta a médico, a astrólogo, a músico, a letrado, y a muchas otras cosas que no son de su profesión' (that a serious monk . . . should deal in medicine, astrology, music, letters, or the many other matters outside his profession)[93] – and that Torres should engage in controversy with a religious. Let each man keep to his proper sphere, is the theme of this paper. So that just as Torres in his Chair of Mathematics and Astronomy at the University of Salamanca should busy himself with his own undergraduates, so should Feijoo confine his activities to matters concerning religion. The inference to be drawn also represents the thinking of the average European scholar of those times, that it is morally and spiritually dangerous to distend orthodox categories of thinking by unorthodox methods of cross-inquiry.

One of Feijoo's admirers, Ernesto Frayer, observed with no surprise in 1727, that Feijoo's supposedly 'sacrilegious' ideas were raising a resistance of hurricane force. And though Frayer rightly mentions jealousy as a contributing factor in antagonism towards Feijoo, this was a type of jealousy realistically related to social anxieties.[94] From within the group where Torres exerted his verbal charm of sophisticated vulgarity a crudely Torresian 'Médico de Sarabillo' exemplifies Frayer's observation. In a *Carta consolatoria* (Consolatory Letter) whose view of the *Teatro crítico universal* is that it might more usefully have served him as lavatory paper in his recent attack of diarrhoea, the Sarabillo physician deals in loud-mouthed indignation with the 'scandal' Feijoo has raised – '. . . el ruido escandaloso que has movido' (the scandalous commotion that you [Feijoo] have roused).[95] But even this abusive critic was not so wildly crude that he was unable to supply a good reason, from the standpoint of his own times, for his spluttering disgust. According to 'herd' understanding, Feijoo, in his medical 'scepticism', had been reasoning outwith those guide-lines of the faith drawn up to preserve the sanctity of Medicine. If Feijoo had so reasoned outwith the guide-lines in purely theological spheres he would, said his Sarabillo critic, have been classed as a Jansenist. The implied threat was serious. And in general the idea that a baffling absence of guide-lines on any subject constituted the greatest threat to religious orthodoxy was an idea no less seriously represented by cartoonists like 'Sarabillo' than by Sentinels of public morals like Araujo.

In an *Antitheatro délphico* of 1727, Zafra Ciscodexa, a milder critic of Martínez and Feijoo, considers their arguments that the early Doctors of the Church must not be expected to have had specialist knowledge, and opposes this argument from the standpoint of established theory. So how does Feijoo know that the Fathers had not studied Medicine? – presumably either naturally or supernaturally[96] – he asks. Feijoo would have difficulty in supplying the concrete evidence to support his own contention, and would find himself in a similarly anomalous position to the one in which he found himself when, on dismissing the medical science of Ramon Lull's *Ars Magna* as valueless, he admitted that he had never read it. Unnecessary as it was, and is, to read Lull's medical science in order to know that it has become outdated, the logic of Feijoo's disparagement was unobvious to all but the minority of students with international outlook.

One of Martínez's more academic critics stung into action by the feckless word 'scepticism' was Juan Martín Lessaca, sometime Professor of Medicine in the University of Alcalá, and, by 1729, appointed Physician to the Dean and Chapter of Toledo Cathedral. In an *Apología escolástica en defensa de la Universidad de Alcalá*,[97] suggestively dedicated to the Cardinal Archbishop of Toledo, he shoulders the task of defending medical orthodoxy against the *Medicina scéptica* and its endorsement by Feijoo. Alcala, the Complutense, like the University of Salamanca, and unlike the Universities of Valencia, Sevilla and certain other outlying places susceptible to foreign influence, was mindful of its traditional rôle as defender of supposedly proved laws of Philosophy and of the unimpeachable quality of their authors and interpreters. Lessaca had read the *Medicina scéptica* as an attack on Complutense curricula and standards. Superficially the *Apología* looks not altogether unreasonable. The Preface, with some degree of justice, attacks the sophistry in Martínez's method of criticism. For the self-confident Martínez sometimes did force general evidence to fit particular cases, causing Lessaca in his turn to answer sophistry with sophistry. Martínez had assumed the need for logic. Can one then know logic, says Lessaca, without Logic? – meaning the standard Logic of Aristotle. Can one 'unteach' without having been taught? Can one, is the implication, then, can one criticize Complutense teaching without having had the benefit of Complutense training? Where, he elaborates in later chapters, where can one learn Medicine if not in a University Faculty? (Chapter III). Why does Martínez assume the name of Hippocrates when Galen, whom he attacks, accepted Hippocrates' doctrine? (Chapter V). What practical knowledge has Feijoo had in the use of quinine and mercury that he should specifically approve of them, or of bleeding and purging that he should condemn them? Therefore, by implication, why cannot Feijoo mind his own business? (Chapter XV). Feijoo, by means of stylistic wit and grace, Lessaca concluded resentfully in Chapter XV, was attracting undue attention to his extramural ideas. But Lessaca on the whole is more cautious than uninformed, and in this respect fairly represents those thoughtful intellectuals who lingered in troubled groups halfway between the comprehensibility of the old and the uncertainty of the new. He recognized that some authorities in the past, Galen, for example, were not incapable of error, yet he contended that their incidental errors left intact the value of their work as a whole. He denied with some truth that universities were ignoring modern discoveries to the extent asserted by Martínez.[98] And, in retaliation to Martínez's statement that all medical theory is untrustworthy, he wondered why scepticism should not be untrustworthy also. Lessaca's main practical objections are made in his first Chapter entitled 'De la secta scéptica', in which he challenges Martínez's belief that human knowledge is acquired not by asserting but by doubting. Probably Lessaca was right to declare himself unconvinced by Martínez's examples of sensory misapprehension and his failure to distinguish clearly between evidence which can justly be obtained from man's sensory impressions, and other kinds of evidence depending on sensory impressions which he should distrust. The fact that Lessaca himself had been encouraged to distrust the findings of Experimental Science by reading contradictory interpretations of such modern discoveries as the circulation of the blood, shows that he was not criticizing without some reflection. It had been assumed in consequence of this discovery, that if blood fails to circulate the patient will die, since the heartbeat depends on the circulation through arteries affecting dilation and contraction.[99] When the Complutense Faculty of Medicine denied that death would necessarily result, said Lessaca, its members were told that they were denying evidence proved by experiment. Yet medical scholars had later confessed that death was not the inevitable result; that the heart's contraction and perhaps dilation is natural to itself and is not dependent on circulation.

Such examples of ambiguity within the limited range of Experimental Science at that stage posed genuine problems over the acceptance of experimental evidence. We now assume that if one experiment seems inconclusive or unconvincing other experiments will be required to correct experiments, until truth is revealed by a process of elimination. Process, like suspension of judgment, or 'unteaching', was a conception difficult for a century newly emerging from the security of dogmatisms to understand. Ambiguity over any given experiment suggested unsoundness not simply of an individual experiment but of the technique of experimentation, especially where unfamiliar instruments were involved. If experiments were proved wrong, or inconclusive, the academic tendency in all but a small minority of scholars was to return to the theoretical reasoning of authorities fully tested, as it seemed, by time, and to right the confusion produced by 'unteaching'.

Lessaca goes more knowledgeably into the details of the general text of *Medicina scéptica* than did Araujo. There are chapters discussing the orthodox interpretation of fevers (Chapters X-XIII) and humours (Chapters VII-VIII). Others in more philosophical areas defend the Schools' use of Dialectics as an exercise in determining the most verisimilitudinous arguments for differing points of view, and imaginatively designate Aristotelian Logic as a guide in 'sol-fa' to Natural Philosophy (Chapters II-IV). Elsewhere Lessaca objects not unfairly to Martínez's bid, understandable if perhaps rather Machiavellian, to enlist the favour of St Augustine, who had spoken of man's ignorance and who, therefore, in Martínez's judgment, had shown himself not averse to 'reformed' sceptics (Chapter II). Certainly Lessaca's overall objection to the *Medicina scéptica* was concerned with its propagation and application of the word *scéptica*, which had generally been accepted as irreligious.

The extensive range of subjects covered by Feijoo's *Teatro crítico universal* inevitably provoked an extensive discussion for and against him over many decades until further criticism was forbidden by the decree of Fernando VI in 1750. He was attacked, sometimes at length, sometimes incidentally, over nearly all his unconventional opinions, from those on Science to those on such general matters as Baroque music in churches.[100] But throughout so many decades the most constant objection to his own and to Martínez's writings was still concerned with their explicit 'scepticism', and the predominant feature of continuing objection was still criticism of that word in isolation. Seldom did their opponents depart from the fashion of ignoring the word's new context. In 1742 a Capuchin Fr. Luis de Flandes,[101] Ex Provincial of Valencia and Murcia, and Inquisitorial Officer for those regions, wrote, so the tone suggests, a genuinely worried commentary in two volumes on the pronouncements of Martínez and Feijoo, particularly opposing Feijoo's dismissal of Ramon Lull as a scientific authority. Unskilled though he seems to have been in his understanding of modern Science and Medicine, he professed a certain respect for Feijoo and Feijoo's intellect, and a certain willingness to accept some of Feijoo's reasoning. His *El antiguo académico contra el moderno scéptico (The old Academician against the modern Sceptic)*[102] deals lavishly with medical and scientific theory in the only language habitual to him – philosophical and religious. But the general theme of his two volumes concerns the danger to the Faith, of which he was an official Guardian, of careless talk in public; and of that propaganda which the masses had not the training to comprehend, and which therefore could only unsettle them. The idea that such risks must be taken in the interests of learning was too new in Spain, as elsewhere, to become quickly acceptable. Its application even in Medicine, the most obvious field of public benefit, would not be taken for granted for several decades more. If men are taught to regard scepticism as an intellectual idea, Flandes asks reasonably enough, will they not eventually bring it to bear on religion and go the way of Luther and English heretics? This preoccupation is at the back of the mind of Flandes' sponsor, Dr Antonio María Herrero, when he speaks of the 'sad consequences' of Feijoo's 'uncertainty'.[103] Flandes elaborated on the same theme. He was willing, he said, to go part of the way towards suspending his judgment in non-religious matters. But, he argued, the very exercise of suspending judgment can be dangerous to the soul: witness the harm done in England by freedom of thought. His explanation would act as a statement of Faith on behalf of the entire college of reluctant anti-enlightenment:

> . . . al cabo de tantos años introducir improporcionadas plantas venidas del Norte, donde los autores viven helados en la Fe y Caridad, y concurriendo todos ellos . . . al desprecio de la Phýsica Pythagórica, de la Lógica Aristoteliana, y de los SS. PP. en cuanto Philósofos, es motivo para recelar, que los herejes con sus halagueñas voces, nos quieren introducir su veneno en la dorada copa de la *experimental Philosofía.*[104]

> (. . . to introduce, after so many years, disproportionate organisms from Northern climes, where the authors' Faith and Charity have been frozen, and where all concur . . . in despising Pythagorean Physics and Aristotelian Logic, and the Holy Fathers as philosophers, is a motive for fearing that the heretics, with their cajoling voices, want to insert into us their poison in the golden goblet of *Experimental Philosophy.*)

For a period still sensitive to the threat of the Reformation it was natural enough for the uninitiated to regard Bacon primarily as a 'first class heretic'.[105] Uninitiated Protestants equally distrusted the scholarship of 'wicked' Catholics and were as vague in their factual knowledge of their activities as was Flandes who knew so little of the English that he named Bacon and 'Francisco Verulamio' as separate examples of heretics.[106]

Altogether, Flandes invites our special interest because he consciously stood at the century's cross-roads, anxiously looking in all directions. He knew less of Science than Torres, but when he weighed up the possible results of disturbed traditions he was more serious, and less selfrighteous than Torres, who always thought from positions of self-interest. Feijoo, of whom Flandes speaks with courtesy, was not, he knew, a 'rigid' sceptic, meaning that he was not in intention or personal practice irreligious. He subscribed to Feijoo's eminence as a writer. But the very force of Feijoo's beguiling originality was likely, he realized, to influence the general reader to imitate him, to use new terms, like his highly publicized 'scepticism', with dangerous ambiguity:

> El daño está en haber el Rmo. Padre atraído al vulgo a que también sean scépticos, pensando tenerle de su parte; y como es más fácil dudar que saber la ostensiva necesidad de la afirmación, o la imposibilidad que induce a la negación, el genio menos cultivado se ha dejado llevar de aparentes discursos, armándose de razones superficiales contra los hombres doctos, hasta imprudentemente gritar contra el honorable gremio de los Médicos, sin poder la indiscreción de los vulgares contenerse, por imaginar ellos haber entendido al Padre Maestro sin sacar otro fruto de sus ingeniosas obras que la detracción, hija siempre de la ignorancia.[107]
>
> (The danger lies in the fact that the Rev. Father has also encouraged the masses to become sceptical, thinking to have them on his side; and as it is easier to doubt than to understand the demonstrative necessity of affirmation, or the impossibility inducing negation, the man of less cultivated talent has allowed himself to be influenced by superficial discourses, arming himself with misleading reasons against the honourable order of Medical doctors, so that mass indiscretion cannot be contained because the masses suppose that they have understood the Father Master, though they have not extracted fruit from his ingenious works other than obloquy, the daughter of ignorance.)

Flandes had rightly observed, within philosophical confines, that much of the difference between Ancients, such as Aristotle and Lull, and Moderns, such as Bacon and Feijoo, had to do with terminology.[108] But in pursuing the thought that anything modern worth saying had already been said more logically by ancient authorities, he missed the point of Feijoo's argument that much of the new terminology was applied to newly observed functions of newly discovered phenomena. For example, that the use of the medical word 'humours', referring to the proportions of earth, fire, air and water in the human make-up was insufficiently complex to account for the entirety of man's physical character; that qualitative definitions, such as calefacient, sarcastically dismissed by Feijoo,[109] though undeniably true so far as Flandes' own knowledge extended, were scientifically inadequate at a period obliged to cater for the mathematical exactitude of Newtonian Physics and medical treatments applied to invisible matter. The change in scientific quarters from descriptively philosophical language to a technical language beyond the understanding of non-specialists was a new fact of intellectual living exceedingly difficult to accept.[110] That the scientist, in his narrow sphere, might pass beyond the deep knowledge of the philosopher, particularly the theological philosopher, so far the highest representative of human knowledge, seemed to contradict the whole meaning of life. So Flandes re-emphasizes in Volume II the consequent danger to the general public, and also to students who would be encouraged by sceptical teachers to suspend their judgment over standard textbooks. One may smile at Flandes' answer to those who asked how, since Lull, for example, had received no scientific education, he could be expected to have a right judgment in all scientific matters, the secrets of transmutation among others. It was, Flandes had answered, by divine intuition. For persons manifestly inspired in religious matters had been considered incapable of making false pronouncements in general. One may smile, yet even in the twentieth century it is not unknown for intelligent persons to ignore the facts of Science beyond their comprehension, and turn at times to the transcendency of faith-healers or reliance on miracles or folk-intuition. It was because the prestige of Science as an independent, self-evident authority was as yet in an early stage of growth that the use of new scientific instruments could logically seem misleading, irrelevant, or valueless. The cautious Flandes was one of those who criticized reliance on man-made instruments for determining what ought, he believed, to be determined by reasoned principles. He thought that such instruments created only confusion; that the very men who put their trust in what they saw through telescopes were the ones who disbelieved in what their eyes saw otherwise.[111] Ultimately the difficulty lay in reluctance to abandon the traditional assumption that particulars must be obtained from general principles, and not general from particulars:

> Además, que la inducción fundada en la experiencia, no puede ser rigurosamente científica, por ser de casos singulares, hasta haberse logrado la de correr todos sus individuos, que son innumerables.[112]
>
> (Besides, induction founded on experience cannot be strictly scientific where based on individual cases, until the experience of all individual cases concerned, and they are innumerable, has been verified.)

From which point of view it was natural to trust an ancient authority like Lull in preference to a modernist like Bacon: a man 'too young in experience', too immature an observer, to dictate to medical scientists and philosophers whose work has withstood the test of two thousand years.[113] To Flandes, Bacon's disagreement with the Ancients was equivalent to a student's disagreement with his Professor.

As for modern instruments, Flandes thought he could exemplify their failings. Thermometers are said never to lie about the degrees of heat and cold. But Flandes was convinced that he had caught a thermometer lying when heat in Murcia one day reached degrees not marked by the instrument.[114] Feijoo, so much given to critical laughter, must have laughed louder than ever at this report. But if the

Murcia thermometer did not show the rise in temperature which Flandes believed he was experiencing, it would truly seem to him, a man used to relying on his own sensations, that thermometers were inaccurate. And if, in fact, the thermometer was shown to be faultily constructed, or accidentally damaged, that would only prove to him its general unreliability. The strange new authority of the thermometer had been provocatively asserted by Feijoo,[115] a gleeful admirer of its powers. Therefore Flandes' own belief in its fallibility and in that of other modern instruments, could lead him only to regret that Feijoo should be meddling in such dubious matters; and that, as Flandes had said in his Dedication to Mañer, another Lullist, members of the Orders should not in any case be indulging in secular activities. The Dedication of Volume II to St Michael the Archangel, was intended to show, through this angelic symbol, that nobody can usurp God's functions by trying to explain the inexplicable. Even though experimentation be a praiseworthy exercise it can produce nothing positive, is Flandes' general belief. It cannot explain cause which is the prerogative of philosophical judgment alone.

As would be expected, Flandes, despite evidence of the latest telescopes, was, for religious reasons, among those who refused to accept the Copernican theory. Since the Church in Spain had not yet pronounced in its favour, even Feijoo in 1726 was as yet unable to accept the theory publicly, though later he declared its validity. But to Flandes this theory, which seemed to contradict the assumption of the Old Testament that the sun moves round the earth, was another symptom of dangerous scepticism calculated to spread confusion and undermine Christian belief by denial of the theological explanation of the universe. To the average guardian of the Faith, as to an average patriot in time of war, blind loyalty was more important for eventual purposes than critical questioning of every command. As the human mind is incapable of comprehending the infinite, its reasoning could be judged as rightfully contenting itself with the knowledge of its limitations and of the dangers of individual speculation. Behind dissention loomed, of course, the spectre of Luther.[116]

There is always a tendency in circumstances of crisis for orthodoxy to be regarded with superstitious awe. When Feijoo ridiculed the orthodox evidence brought forward to prove that the world is perfect in form because it consists of three dimensions, Flandes reminded him, with a reference to Lull, that three, the number of the Trinity, is a sacred symbol, and that the perfect, three dimensional form of the world is consequently related to the Godhead.[117] Should anyone now find difficulty in understanding the way in which religious awe towards an accredited religious authority, such as Lull, could deteriorate into superstition, he has only to survey the modern persistence of Astrological literature or observe the supernatural significance still attached by the average person to the number thirteen.

Incidentally, the supposed orthodoxy of established criteria in all subjects had, before the publication of Martínez's organized doubts, come into conflict with new interpretations of the very word *criticism*. In 1720 a worried Cristóbal Fuertes y Núñez ('Racionero de la Iglesia Metropolitana de Zaragoza') (Prebendary of the Metropolitan Church of Zaragoza) had written a *Breve desengaño crítico de la historia de España del Doctor Juan de Ferreras (Short Critical Disillusionment about the History of Spain of Dr Juan de Ferreras)*. Here, while defending the rights of scholars to view evidence critically, and while dissociating himself from the *vulgo*'s tendency to treat all criticism as scandalous, he is distressed at what seems to be a forlorn prospect of criticism openended. Academic reasoning had been guided by rules. His cry is for guide-rules of criticism:

> Sería temeridad culpable el no confesar las conveniencias del buen uso de la *Crítica*, cuando sin ella apenas podrá distinguirse lo falso de lo verdadero . . . Todo lo que no se mide por sus Reglas se tiene por inconstante; pero la mayor inconstancia se ha descubierto, en que no tiene reglas ciertas la *Crítica*; pues en cada autor son arbitrarias. Los que se acreditan por de mayor carácter en la República de las letras, no han convenido aún en los principios del Arte, y siendo éstos, precisos en cada licencia, y que por ser de notoria verdad están fuera de disputa, en la Crítica no hay quien no dude en todo, o en parte, de éstos que suponen por principios . . . Llenos están los libros críticos de inútiles controversias, que dejarían de serlo, si hubiera reglas, y principios para determinarlas; pero se ha hecho país libre la crítica, donde cada uno camina según su fantasía, abusando de la licencia que da la libertad, para hacer juicio . . .[118]

> (It would be culpable temerity not to admit the convenience of the good use of the *Critique*, when without it one could hardly distinguish the false from the true. Everything not measured by its rules is considered to be inconstant; but the greatest inconstancy is revealed in that the *Critique* has no fixed rules, for in each author they are arbitrary. Those accredited of greatest consequence in the Republic of Letters have not yet agreed about the principles of Art, and though such principles are requisite in every profession, and being of authorized truth are beyond dispute, yet in the *Critique* there is nobody who does not have doubts in general or in particular about these supposed principles. Critical books are full of useless controversies which would cease to be controversies if rules were to exist, and if there were principles to determine rules, but criticism has turned into open country where everyone walks according to his fancy, abusing the licence given by such freedom to make judgments.)

Ferreras' work to his mind was a case in point. For the historian had dismissed as unlikely much evidence that bore quantitative weight and the weight of ancient tradition – such as the 'evidence' of Simón Metaphraste that St Peter had journeyed to Spain: the evidence, Ferreras thought, of an *orador* (preacher) rather than of an *historiador* (historian). 'Rules' deriving from quantity and antiquity of sources would continue to seem reasonable even to intelligent and responsibly minded scholars. Exceptional adventures into uncontrolled criticism, that is, criticism outwith current definition, would long continue to suggest wild guesswork and irresponsibility. For this very reason there were many who approved of much of Feijoo's freelance researches only up to, and not beyond the point of his freelance interpretations of them. Felipe Brizeño y Zúñiga goes so far as to say that 'querer arrancar del vulgo la vulgaridad, es querer arrancar el vulgo del vulgo' (to want to tear mass-thinking from the masses is to want to tear the masses from the masses). But, depending on the only rules of criticism natural to his mind, he can object that Feijoo should treat so many *comunes opiniones* as *errores comunes* (common opinions as common errors):

> Lo que es de todas maneras intolerable, es, que a las comunes opiniones que estriban en la autoridad de graves Historiadores, diligentes Geógraphos, y Sabios Naturalistas, las infame el P.M. con la nota de errores comunes. Sea sufrible el que las impugne, no que las degrade . . .[119]
>
> (What is in every respect intolerable is that the common opinions supported by the authority of serious historians, diligent geographers, and wise naturalists, should be defamed by the Father Master with the stricture of common errors. To contradict [such opinions] may be acceptable but not to degrade them.)

To explore the unknown without the aid of a reputable handbook of directions was to sail on the open sea in a boat without a rudder. None but the unexceptional adventurer could be expected to think otherwise. As a parodying 'crítico moderno' (modern critic) of Ferreras' technique put it:

> Enseño a no creer;
> quien creer no quiera,
> (sea lo que fuere)
> acuda a mi escuela . . .[120]
>
> (I teach disbelief;
> he who may not want to believe,
> (whatever the matter)
> let him hasten to my school . . .)

Openended 'criticism' in general, like medical 'scepticism' in particular, would inevitably alert serious minds to serious dangers.

Drifting into the 'sixties, the prevailing distrust of religious and academic permissiveness continued to propagate. In 1761 the *Duende especulativo*,[121] which fancied itself as a kind of Spanish *Spectator*, could complain that every new discovery or invention had met opposition from dogmatists, traditionalists, and conscientious objectors; that the names of foreigners such as Newton and Wolf were still suspect; that Classical and Scriptural backing was still deemed necessary for every mental enterprise; that 'an inveterate preoccupation still tiranizes man's understanding'.[122] Nor was the *Duende* itself so openminded or unshockable that it could countenance the entry of criticism through Church doors to investigate even the stylistic quality of preaching. Evidently Isla's *Fray Gerundio* of 1758 seemed to the Duende to overstep bounds of legitimacy.[123] Nevertheless, climatic changes were on the way. No longer were arguments over points of Science and Religion restricted largely to academic quarters. Thanks to the imaginative vitality of Feijoo and Torres, especially when they wrote at cross-purposes, thanks, too, to their impassioned defenders and detractors, the subjects of debate, and fears involved in these, had become widely advertized: a preliminary stage in the development of tolerance.

By the time the journalist Francisco Mariano Nipho, in 1765, in a pre-notice of his book *El maestro del público para padres de familias y públicos maestros*, defended Experimental Science in Feijoon strain against the prejudice which 'nos hace los sujetos más ridículos de la Europa, y el primer objeto de sus sátiras' (makes us the most ridiculous people in Europe, and the prime object of its satires),[124] his general public was too familiar with the call for re-education to be unduly shocked or indignant. The need for Hippocrates or Aquinas to come dialectically to the aid of the enlightened was less imperative. Familiarity with the principle of *Epoche*, even for purposes of disagreement, was a state of mind in their readers on which earlier writers and their colleagues in outlying universities could not depend. Therefore we should not lightly dismiss the value of apparently time-wasting polemic on the part of those incapable of understanding scientific evidence. In mid-century the theologian Francisco Soto y Marne is typical of many who still treated the word with horror. Yet his *Reflexiones crítico-apologéticas sobre las obras del R.P.M. Feijoo*, which discusses Feijoo's treatment of Ramon Lull, enlarges probably the more vehemently on the dangers attached to 'scepticism' in that it was becoming, as a general idea, more widely circulated. His concern is the common concern about the extension of free thinking to religious areas and the attraction of innovation:

> . . . aquellos filósofos, que enamorados de la novedad, o dominados de un scepticismo caprichoso, desprecian el sentir de los antiguos, entregando, con terca tenacidad, su asenso, a la obsesada voluntariedad de su presuntuoso discurso, porque habituados estos ingenios a discurrir con libertad por la esfera de la naturaleza, introducen este pernicioso libertinaje en el sagrado, hemisferio de la Gracia.[125]

> (. . . those philosophers who, enamoured of novelty, or dominated by a capricious scepticism, despise the judgments of the Ancients, letting loose their assent, with obstinate tenacity, over the obsessive free range of their

presumptuous discourse – because such geniuses, accustomed to ramble freely throughout the sphere of nature, introduce this pernicious licentiousness into the sacred hemisphere of Grace.)

However, familiarity with alarming concepts tends to breed, if not acceptance, at least some degree of indifference. Twentieth-century acceptance or indifference to once startling facts and judgments has come about similarly because the concepts involved have been vulgarized. If familiarity does not lead necessarily to understanding, it commonly leads unconsciously to suspension of judgment. Even the noisy popularity of Torres' Prognostications, which the *Duende* viewed with disgust, proved a means of drawing attention to everything else that Torres wrote, including his own acceptance or rejection of rationalistic tenets. Which does not mean that either Torres or his readers had been driven to fully reasoned conclusions. Torres' fits of reason or emotion, or of mixtures of both, were determined by any mood of any moment as well as by his current prejudices. Rather was it that the very hypnotism of repetition had caused them all to assimilate painlessly many notions about change and its insecurity which had seemed impossible to be swallowed with deliberation.[126]

NOTES

1 Peset, *op.cit.*, pp. 29, 31, 42, 172.

2 *Ibid.*, p. 329. The author mentions that, among books ordered, were 83 volumes of the records of the Royal Academy of Sciences of Paris.

3 *Ibid.*, p. 324.

4 *Medicina scéptica y cirugía moderna*, 2 vols. (Madrid: n.pub., 1722).

5 Second edition, Madrid: Jerónimo Rojo, 1727. Feijoo's *Apología* is dated 1725. A third edition was published in 1748.

6 See, for example, his *La Real Academia Sevillana de Buenas Letras en el siglo XVIII* (Madrid: C.S.I.C., 1966).

7 *Medicina scéptica*, Vol. I, 1722.

8 See Bernardo López Aranjo y Ascarraga, *Centinela médico-artistotélica contra scépticos* (Madrid: n.pub., 1725), pp. 71ff., and see pp. 118-33 below.

9 *Aprobación* to the edition of 1722.

10 See John Wesley, *Journal* (London and New York: Everyman, 1930), Vol. I, p. 500. Compare misunderstandings over Dominican terminology referred to in a Papal Decree of 1724. It is translated for the *Gaceta de Salamanca*, of Jan. 16, 1725 (B.Nac., R. 23981).

11 *Medicina scéptica*, *ed.cit.*

12 *Ibid.*

13 *Op.cit.* See Martínez's Prólogo. He takes the words, he says, from Chrysostom.

14 *Op.cit.*, Introducción.

15 *Op.cit.*, Prólogo.

16 *Op.cit.*, Introducción.

17 See, for example, the *Memoirs* of Edward Gibbon on the prejudices of Oxford University (London: Everyman, 1948), pp. 49ff, *et passim*.

18 I have used the convenient modern edition published by Macdonald, London and New York, 1971. See the Introduction to this edition by Lester S. King, pp. x ff.

19 See Brian Inglis, *A History of Medicine* (London: Weidenfeld and Nicolson, 1965), pp. 108ff.

20 *Medicina scéptica*, Vol. II (Madrid, 1725), p. 75.

21 *Op.cit.*, Vol. I, pp. 79, 95ff.

22 *Teatro crítico universal*, VIII, Discurso X, *ed.cit.*, p. 208.

23 See Everyman ed., London, 1932, Vol. I, Ch. XIII, pp. 286ff.

24 On this subject see Daniel R. Reedy, *The Poetic Art of Juan del Valle Caviedes* (Chapel Hill, N.C.: Univ. of North Carolina Press, 1964), Ch. V, pp. 60ff.

25 *Los sueños*, ed. J. Cejador y Frauca. Clásicos Castellanos (Madrid: Espasa Calpe, 1954), pp. 32, 36, 46, 81, 134, 200ff., 214; Vol. II (1960) p. 83, etc.

26 See n.8, above.

27 *Centinela . . ., ed. cit.*, p. 301.

28 *Op.cit.*, p. 10.

29 See the *Aprobación* of Rmo. P.M. Manuel Irigoyen.

30 *Centinela . . .*, p. 10.

31 *Op.cit.*, p. 51.

32 *Op.cit. Aprobación* of Rmo. P.M. Fr. Mathías Antonio Navarro y Aguilar.

33 See Feijoo's 'Aprobación apologética' to the 1727 edition of the *Medicina scéptica*. It is dated Oviedo, Sept. 1st, 1725.

34 *Medicina scéptica*, Vol. I. See Conversations 7, 9, 14-24, etc., and see pp. 18-33 for discussion of medical details.

35 See, for example, *op.cit.*, II. Conversations 26, 27, *et passim*.

36 *Centinela . . ., ed.cit., Aprobación.*

37 See, for example, Peset, *op.cit.*, p. 359.

38 See John Wesley, *Journal, ed.cit.*, Vol. I. Nov. 24th, 1739; Dec. 13th-14th, 1739; *et passim*.

39 *Centinela . . .*, p. 86.

40 *Op.cit.*, p. 248.

41 *Op.cit.*, p. 322.

42 See Feijoo, *Ilustración apologética al primero y segundo tomo del Teatro crítico* (Madrid, 1729), p. 183; and see I. L. McClelland, *Benito Jerónimo Feijoo, ed.cit.*, p. 118.

43 See for example, Nicole Rochaix, 'José Climent et la lutte contre l'ignorance dans l'Espagne du XVIII siècle', *Les Langues Néo-Latines*, LXXII (1978), No. 1, 26-64. See p. 34 *et passim*.

44 *Centinela . . .*, p. 336.

45 *Op.cit.*, p. 341.

46 *Medicina scéptica*, Vol. II, 1725.

47 See McClelland, *Benito Jerónimo Feijoo, ed.cit.*, pp. 57ff.

48 *Fray Gerundio*. Clásicos Castellanos, Vol. II, (149), p. 18.

49 See *Centinela . . .*, p. 246, and *Medicina scéptica* II, *ed.cit.*, p. 33.

50 See, for example, Martínez's *Carta defensiva que sobre el primer tomo del Theatro Crítico Universal . . . le escribió su más aficionado amigo . . .* (Madrid: Imprenta Real, 1726).

51 Torres, *Entierro del juicio final y vivificación de la astrología . . ., Obras completas* (Salamanca, 1752), Vol. X. See pp. 140ff. It was written in 1727.

52 See, for example, Feijoo, *Cartas eruditas (Madrid, 1742-60)*, vol. III, C. 16; vol. IV, C. 26.

53 *Medicina scéptica*, II, *ed.cit.*, p. 84.
54 *Teatro crítico universal* (Madrid, 1736), Vol. VII, 13, pp. 325-327.
55 *Medicina scéptica*, Vol. II, *ed.cit.*, Conversation 27, p. 37.
56 *Op.cit.*, C.26, p. 15, *et passim.*
57 *Op.cit.*, C.27, p. 32.
58 *Ibid.*
59 *Op.cit.*, C.29, pp. 77ff., C.31, pp. 108-110.
60 *Op.cit.*, C.26, p. 26.
61 *Op.cit.*, C.26 and C.29.
62 *Op.cit.*, C.25, p. 6.
63 *Op.cit.*, C.26, p. 12.
64 Feijoo, *Cartas eruditas*, Vol. III, C.4, p.30.
65 'Aprobación apologética' to *Medicina scéptica.* There is no pagination. For modern research on the dating of the *Aprobación apologética* and its relation to the earliest publication of Feijoo, see P. Álvarez de Miranda, 'La fecha de publicación del primer escrito de Feijoo', *Dieciocho*, 9 (1986), Nos. 1-2, 24-34.
66 See Feijoo, 'De lo que sobra y falta en la enseñanza de la medicina', *Teatro crítico universal*, VII, 14, and Martínez, *Medicina scéptica*, II, C.31: 'Que la lógica artificial es del todo inútil para la medicina'.
67 'Aprobación apologética'. See I. L. McClelland, 'The Significance of Feijoo's Regard for Francis Bacon', *Studium Ovetense*, I (1976), 249ff.
68 Torres, *Visiones y visitas . . .*, ed. Russell P. Sebold, Clásicos Castellanos 161 (Madrid: Espasa Calpe, 1966), pp. 14, 120, 246. See I. L. McClelland, *Diego de Torres Villarroel* (New York: Twayne, 1976), pp. 106ff.
69 Torres, *op.cit.*, p. 250.
70 Quevedo, *Los sueños*, *ed. cit.*, I (Clásicos Castellanos 31), pp. 29ff.
71 See, for example, Torres' *Correo del otro mundo al Gran Piscator de Salamanca* (Salamanca, 1725), pp. 28-29, *et passim.*
72 See Chapter 4 below.
73 Isla, *Glosas interlineales . . .* See *Colección de papeles crítico-apologéticos que en su juventud escribió el P. Joseph Francisco de Isla* (Madrid, 1788), p. 85. (British Library).
74 See, for example, Torres, *Posdatas a Martínez* (Salamanca, 1726); *Entierro del juicio final. . .* (1727), *Obras completas*, *ed.cit.*, Vol. X, etc. On Torres' supposed empiricism see R.P. Sebold's Introduction to the annotated edition of Torres' *Visiones y visitas* (Madrid: Espasa Calpe, 1966).
75 Feijoo, *Teatro crítico universal*, I.8.
76 Torres, *Posdatas a Martínez . . .* (Salamanca, 1726), pp. 11-12. It is contained in a collection of papers by Torres in the Biblioteca Nacional.
77 Torres, see *Obras*, 1752, Vol. I, Preface.
78 Torres, *Vida*, CC., pp. 104-05.
79 It was published in Madrid. The work itself is undated, but it bears a *censura* of 1727. (B. Nac.).
80 *Escuela médica . . .* (Madrid, 1727), Aprobación by D. Antonio Fernández Lozoya.
81 *Ibid.*
82 See McClelland, *Benito Jerónimo Feijoo*, pp. 67ff.
83 Ribera, *op.cit.*, Preface.
84 'Abuso de las disputas verbales', *Teatro crítico universal*, VIII, 1.
85 Ribera, *op.cit.*, Book II, p. 154.
86 The argument is discussed in Ribera, *op.cit.*, Book II, Chapter 2.
87 Ribera, *op.cit.*, Book II, p. 171.

88 These normally rational theologians were, of course, aware of the dangers of encouraging the public to extend miracle-thinking simplistically. See, for example, J. Wesley, *Journal*, Vol. II, Everyman, pp. 180-81.

89 See, for example, M. Martínez, *Juicio final de la Astrología* (Madrid and Sevilla, nd.), *passim*.

90 *Escuela médica . . ., ed.cit.*, Chapter 6, p. 204.

91 *Medicina cortesana, Satisfactoria . . . en respuesta a la honoratísima carta que . . . Feijoo escribió al autor . . .* (Madrid, 1726), p. 114.

92 See Pedro Aquenza, *Breves apuntamientos en defensa de la medicina, y de los médicos, contra el Theatro Crítico Universal* (Madrid, Licences, 1726), pp. 1-2. (B. Nac.).

93 *Carta de Pascuas, que desde Guadalcanal escribe un Barbero a Don Pedro del Parral, vecino de Madrid, diciéndole lo mal que le han parecido los papeletos del Rmo. P. Feijoo, de Torres, de Aquenza, de Martínez, de Ribera y del Músico* etc., (Dec. 12, 1726), p. 2.

94 *Discurso filológico crítico, sobre el corolario del Discurso XV del Theatro crítico universal* (Madrid, 1727). (British Library).

95 *Carta consolatoria del médico de Sarabillo a un discípulo suyo, sobre las inquietudes que ha movido el Teatro crítico, que ha sacado a luz el P.M. Fray Benito Feijoo, y advertencias médico-teológicas a dicho Padre.* There is no place or date of publication. A reference on p. 2 mentions 'last year, 1726'. See p. 13 (British Library).

96 See *Antitheatro Délphico judicial jocoserio, al Theatro Crítico Universal del R . . . Feijoo* (Madrid, 1727) (British Library).

97 *Apología escolástica en defensa de la Universidad de Alcalá, y demás universidades de España. Contra la 'Medicina Scéptica' del Dr. Martínez. Respuesta al Discurso de la Medicina del 'Teatro crítico universal'* (Madrid, 1729). (Biblioteca Nacional, Madrid).

98 Compare Peset, *op.cit.*, pp. 251ff., *et passim*. See also: Mariano Peset and José Luis Peset, *Gregorio Mayáns y la reforma universitaria* (Valencia: Publicaciones del Ayuntamiento de Oliva, 1975), Serie Menor II.

99 Compare F. Hoffmann, *op.cit.*, pp. 13ff.

100 See, for instance, Eustaquio Cerbellón de la Vera, *Diálogo armónico sobre el Teatro crítico universal: en defensa de la música de los templos* (Madrid, 1726) (British Library).

101 In relation to Feijoo's confusion over the identity of Flandes, see Peset, *op.cit.*, pp. 405ff.

102 *El antiguo académico contra el moderno scéptico, o dudoso, rígido o moderado. Defensa de las Ciencias y especialmente de la Phýsica Pytagórica, y Médica en el conocimiento y práctica de los Médicos Sabios*, Vol. I (Madrid, 1742); Vol. II (Madrid, 1744). (Biblioteca Nacional, Madrid).

103 *Op.cit.*, Vol. I, Aprobación.

104 *Op.cit.*, Vol. I, p. 39.

105 *Op.cit.*, Vol. I, p. 72.

106 *Op.cit.*, Vol. I, Discurso 5.

107 *Op.cit.*, Vol. II, pp. 5-6.

108 *Op.cit.*, Vol. II, p. 73.

109 See p. 29 above.

110 For an example of qualitative arguing see Flandes, *op.cit.*, Vol. I, pp. 181-82, where he explains the working of a magnet, whose iron has been forged into a drier substance than the earth-material from which it originated, as desiring to 'reducir a la naturaleza de piedra, que fue su origen lo que tiene recibido artificialmente como superfluo' (to reduce to the nature of stone, which was its origin, what it has

artificially received as a superfluous addition). Flandes will presumably be influenced by Hippocratic principles such as: '. . .man is not a unity but each of the elements contributing to his formation preserves in the body the power which it contributed. It also follows that each of the elements must return to its original nature when the body dies; the wet to the wet, the dry to the dry, the hot to the hot and the cold to the cold. The constitution of animals is similar and of everything else too'. See *Hippocratic Writings* (Harmondsworth: Penguin Classics, 1983), p. 262.

111 *Op.cit.*, Vol. I. See p. 73.

112 *Op.cit.*, Vol. I. See pp. 85-86.

113 *Ibid.*

114 *Op.cit.*, Vol. I, p. 158.

115 See, for example, Feijoo, *Teatro crítico universal*, II, 13.

116 See Flandes, *op.cit.*, Vol. II, Discursos, 3, 4, 10.

117 See Flandes, *op.cit.*, Vol. II, p. 109.

118 C. Fuertes y Núñez, *Breve desengaño crítico de la historia de España del Doctor Juan de Ferreras* (Zaragoza, 1720), pp. 4-6.

119 Felipe Brizeño y Zúñiga, *Juicio particular del juicio universal* . . . (Madrid: Ant. Marín, 1728), p. 7.

120 *El crítico moderno, transformado en corredor de oreja*, n.d., n.p. (Biblioteca Nacional, Madrid).

121 *El duende especulativo, sobre la vida civil, dispuesto por D. Juan Antonio Mercadal* (Madrid, 1761). The Biblioteca Nacional has 17 numbers: June 9-September 26, 1761. (R.21084).

122 *Op.cit.*, June 19, 1761, p. 90.

123 See *op.cit.*, June 13, 1761, p. 14, and pp. 97ff. *et passim* below.

124 The notice, of six pages, is entitled 'A todas las personas de sano corazón, que desean ardientemente la felicidad de la Patria, mediante la buena, y oportuna educación de la juventud de uno y otro sexo, para gloria y fortuna del estado'. It is dated Madrid, April 23, 1765 (B. Nac. R.23981).

125 *Reflexiones crítico-apologéticas sobre las obras del R.P.M. Feijoo* (Salamanca, 1748-49), 2 vols. Vol. I, pp. 16-17.

126 For other examples of polemic involving particularly Martínez and Feijoo see McClelland, 'Feijoo in Polemic' in *Benito Jerónimo Feijoo*, pp. 107ff.

CHAPTER 3

The *Vulgo*-Conception of Scientific Evidence

I General Science

Ironies of terminological misunderstanding in the fallow mind of the century were not confined to definitions of the word 'sceptic'. Fallowness, by its very state of multiple receptivity, is open to a confusion of ideas. Certain other terms which became liable to misinterpretation in current usage and which soured debate were, for instance, ones referring to the nature of scientific evidence. Among notably misleading adjectives paraded with theatrical confidence by the folk-scientist Torres Villarroel and his peers were 'demonstrable', 'practical' and 'experimental'.[1] They were words used to indicate conclusively physical proof as obtained through normal operations of human senses which had generally been regarded as the ultimate means of testing phenomena. Subjective conceptions of demonstrable evidence were part of a mind-pattern created by the needs of those who tried to contribute to fashionably modern logic by combining certain selected principles of old philosophy with selected principles of new rationalism. It was a pattern shaping itself in many scientific disciplines, including Medicine and the related 'Astrological Astronomy', which in the variety of its implications greatly contributes to our understanding of the troubled *vulgo*-mind. For this reason of technique, and of the authors' potentialities for change, and not because of the intrinsic value of typical expressions of philosophical compromise, it is helpful to follow in different disciplines some of the arguments concerned.

Torres himself moved mentally in a grey academic area, somewhere between the ground occupied by empiric specialists and that inhabited by dogmatic theorists. It was a wide, uncomfortably exposed area in the experience of the average student of Science – especially in medical Science – who was disturbed by new threats to established theory, yet who was not unaffected or uninfluenced by certain features of inductive propaganda. Such men include the aforementioned opponents of Martínez's 'scepticism', Bernardo López de Araujo, Pedro Aquenza, the 'médico de Sarabillo', Francisco Suárez de Ribera, Juan Martín de Lessaca, who wrote on medical subjects between 1726 and 1729, and many other writers on medical and general scientific matters in the first half of the century.[2] For present purposes of assessing generalized conceptions of 'scientific evidence', these scholars may be adequately represented by their most conspicuously self-expressive colleague, Torres the 'Know-all',[3] a typical spokesman for those who understood experimental Science in part and misinterpreted it as a whole: a popular writer, creatively imaginative, and so a spokesman in a peculiarly advantageous position to communicate his half-way understanding of enlightened Science to the general public of his times, and to lead modern readers to an appreciation of eighteenth-century perplexities.

The subject matter of his treatises and polemical essays was largely determined by the latest arguments of other writers. Feijoo's *Teatro crítico universal*, more punctilious in reference and documentation than Torres' works, discussed all the current intellectual questions and was inspired, in its turn, by scholarly discussions abroad. At the time when Torres began to write, the *Journal des Savants* and the *Mémoires pour L'Histoire des Sciences et des Beaux Arts* of Trévoux had already called European attention to their discussions on new medical attitudes to blood-letting and purging, and on the treatment of fevers and venereal diseases. They had publicized the significance for diagnostic purposes of the circulation of the blood. In the first volume of the *Teatro crítico* Feijoo had broached the subject of blood-transfusion and blood-groups. From the 'thirties onwards new medical books, and from the late 'forties, foreign works on electricity, and reports on English and French experiments in Physics, figured largely in Spanish Press-advertisements. In the 'fifties a medical body in Madrid, the Royal Association of Our Lady of Hope, began to announce awards for answers to medical and scientific questions. Accounts appear in the Press of physical abnormalities in nature and unusual case-histories. When gauging the mental atmosphere of his Spanish times it is helpful to listen to Torres' voice on such subjects, especially in its emotional choice of material, in its undiscriminating emphasis, and in its picturesque folk-jargon, which, however artistically contrived, could, for scientific purposes, be impressively and therefore influentially beside the point.

It may be said that his claim to the 'demonstrable' evidence of his assertions became the most prominent feature of his scientific reasoning and that the importance he gave to the principle of demonstrating truth practically was a genuine sign of his rationalist times. Yet, never, for all his profession of intellectual humility, did his pride allow him to learn substantially from free-range rationalists outside his immediate experience, and his conception of demonstrable evidence remained over-simplified, restricted to the idea of assembling and classifying rather than to that of dissecting and analysing. It was important to Torres, who saw himself as a leader of modern scholarship, to profess enlightened ideas and to display his acquaintance with famous international names. In practice, however, he tried to follow the inductive way by processes of deduction. 'Natural reason' and the 'experience of the ages', Torres' fictitious peasant of the *Cartilla rústica* was told in 1727, are the best guides to that practical understanding of man, in which the scholar is so deficient.[4] This high-sounding echo of Renaissance Philosophy was, and still is, all very well so long as the findings of subjective reasoning coincide with those of scientific objectivity, as indeed they often do. Beyond that point, however, Torres did not progress. He never quite appreciated the fact – which his century in every country found difficult to assimilate – that new instruments and machines, especially the microscope, thermometer, barometer, air-pump, and, later, rudimentary conveyers of electricity, by revealing more than a man's five senses can reveal, and so by proving hitherto unimaginable truths, were rendering natural reason inadequate for material purposes. The wisest of his contemporaries were gradually disengaging the Sciences from old commitments with Philosophy and Theology. But Torres' religion was less sophisticated than that of some of his Benedictine and Jesuit contemporaries, and more influenced by folk-prejudice. Whenever he came to a religious turn in any argument he ceased to be openminded, not so much out of loyalty and considered conviction, as out of habit.

For example, when replying to his critics on the 'evidence' of astrological influences on the affairs of men, Torres usually missed the material point. Can anybody doubt mathematically proved demonstration? is his challenge, when referring to calculations relating star-positions to men's affairs, in his *Desprecios prácticos . . . a los prácticos avisos* of 1725.[5] 'I am a stern doubter [in Science], I believe nobody who cannot come to me with a practical demonstration' [of his theories], he repeated grandly in the same year, in the *Correo del otro mundo*,[6] when speaking of the medical effects of bloodletting. 'The fact that Galen advocated bloodletting', went on Torres, accenting his own empiricism, 'does not mean that if I observe bad effects from it I must argue that bloodletting is always advisable.' On the other hand, Torres' doubts here did not mean that he was prepared to revise ancient theories of Galen, Hippocrates or anyone else, least of all on his pet subject of star-influences, unless such revision suited the convenience of his current argument. In his *Posdatas a Martínez* of 1726 he is to be heard proving the physical effects on man of eclipses and comets by saying that their bad influence has always been recognized as a 'most certain principle' by ancient philosophers. And he blames Martínez for daring to disagree with Galen and Hippocrates and for neglecting what Torres regards as the evidence of the senses authorized by ancient systems of Philosophy. His question: '. . . If the moon, sun, planet, air, comets, do not [according to Martínez] account for human illness, what else, I should like to know, could account for it?'[7], was meant as a shattering retort. But its chief effect must have been to confirm the Feijooists in their low estimate of herd-mindedness. Folk-reasoning, personal inconsistency, the need, common to all save rare thinkers, to preserve a protective human link between Science and Philosophy, explain most of Torres' outdated arguments. To modern ears his talk in, for example, *Entierro del Juicio Final*, of dangers to the sick when the course of their diseases reaches seven and multiples of seven days, because God's creation was the work of six natural days, with a break from the regulated course of events on the seventh, will sound simplistic. So will his insistence in the same article that there can be only seven planets because no more have been seen; or his explanation in *Juicio y pronóstico del nuevo cometa*, of 1744, that the comet might, on this unique occasion, be beneficial because of its height, transparency and proximity to purer areas than earth. As motions of Scholastic debate, his talk of this nature might equally well be arguable in reverse. But these were the ideas of the average intellectual in all countries even in the Age of Reason. Torres was no more backward in his thinking than the majority of European academicians and no less puzzled or uncertain. His remarks on the benign comet would be in part a response, unconscious in all likelihood, to growing criticism of unscientific explanations of comets. At the same time they would represent to his doom-mongering imitators in prediction a triumph of artistic individualism. For, by 1744, the year of the new comet, after the subject of sky-phenomena had been well aired by Feijoo, anti-Feijooists, and astrological prophets in sensational competition, Torres was only one of many Spanish interpreters of sky-signs, and not one of the most disinterested.

As a disputant, Torres was the child of his disputatious, rationalistic times, and the criticism directed against almost all of his controversial publications by declared, disguised, or anonymous authors, nearly matched in quantity and force the criticism massed against Feijoo. Some opponents were erudite and enlightened observers of foreign scholarship. Some, like Torres himself, enjoyed an ingenious fight for ingenuity's sake. Some were petty scribblers deserving of a little of his big contempt, especially over literary matters.[8] Some disputed specific points of fact. Some objected to his imperious bragging in the cause of humility. Some justifiably disliked his superficiality. Torres 'has written on all kinds of subjects without knowing what he was writing about',[9] said Juan Antonio Mariscal y

Cruz. 'Torres knows a lot of things badly', said Julián Rodríguez Espartero.[10] If Torres' pious advice were taught by his example, said Mañer, it might do more good.[11] The *Diario de los literatos*, while noting, somewhat maliciously, Torres' borrowings or even 'robos' (thefts) of phrases from Quevedo, treats him lightheartedly as the scholars' jester:

> No solamente los literatos han hecho su delicia de la lectura de sus obras; también los hombres doctos han descansado de la tarea de estudios más severos, solicitando lograr en ellas algunos festivos intervalos: no se ha usado de más poderoso exorcismo para lanzar el demonio de la melancolía.[12]
>
> (It is not merely the [ordinary] lettered men who have taken delight in reading his works; very learned men also have rested from the task of most serious studies by seeking to obtain in them some festive intervals: no more powerful exorcism has [ever] been used to cast out the demon of melancholy.)

While the *Resurrección del Diario* . . ., which, stating that 'también en los literatos hay Vulgo (y no corto)' (among intellectuals too, there is a common herd, and not a small one), chastised Torres for his scientific pretensions:

> Todo él [*i.e.* the contents of his article on earthquakes and the relationship of the earth to the human body] es un mero vejestorio: una especie de cecina rancia, que equivocadamente creyó Torres preservarla con la sal corrompida de su gracejo . . .[13]
>
> (All of this is something gone stale: a kind of cured meat turned rancid which Torres wrongly thought to preserve with the unsavoured salt of his wit.)

Some critics attacked him on all of these scores. To take specific objections first, probably his most formidably personal adversary, for Feijoo and even Martínez dealt in ideas rather than in personalities, was the sophisticated Father Isla, whose attacks on Torres, under a variety of pseudonyms, understandably angered that captious Astrologer intolerably. For us it is easy to sympathize with both contenders: with Isla for his defence of scientific enlightenment; with Torres for his personal defence of his inner uncertainties, and for the compelling fantasy with which he wrapped these up. At the time, however, it was the words of debate, and not remote reasons, which caught public attention. In 1726, Isla anonymously attacking the Salamancan 'Stargazerlets', 'Predictionlets' and 'Postscriptlets', in imitation of Torres' style,[14] objected specifically to confusions of date in the Astrologer's calendar-making, and to his reliance on Ptolemaic Astronomy, underlined also by Martínez. In accordance with latest findings, he challenged Torres' belief that comets were made of terrestrial material, and referred him to less outdated scholars than Ptolemy: to Galileo, Copernicus, Gassendi, Descartes . . .

Isla was joined in his attack on inaccuracy by other critics also. An anonymous writer in a *Respuesta del vulgo a D. Diego*, of 1744, argued that Torres' explanation of recent sky-phenomena as a result of the predisposing weather of previous months, could hardly be acceptable when his account of the weather over that period was only partially correct.

Several critics, including Isla and the Juan Antonio Mariscal y Cruz of *Consejos amigables a Don Diego*, in 1728, tried to disillusion Torres' admirers about the originality of his astrological technique, by stating that he had merely copied from ancient sources. Martínez in the *Juicio Final de la Astrología*[15] contradicted Torres' reading of Hippocrates who, according to Don Diego, spoke of the influence of the stars on health, but, according to Martínez, merely on the effect on health of the different seasons: a line of discussion which Torres and his opponents pursued untiringly. Salvador Joseph Mañer, in 1728, commented astringently, among many other astringent observations, that modern telescopes proved comets to be connected not with earth but with the stars. Like Isla, Mañer had seized the opportunity in this *Repaso general de todos los escritos de . . . Torres* fairly and unfairly to destroy Torres' image of himself in all particulars, and was accompanied in this general endeavour by a host of other detractors. In summary, however, it should be said that probably the most apposite remark from Torres' critics came from a 'Don Juan de Quevedo', allegedly a Professor at Salamanca, in a *Pepitoria crítica* of 1730. Here, speaking of his colleague's attacks on Feijoo and Martínez, he claimed that Torres had not read Feijoo's *Teatro crítico* properly.[16] The observation is thoroughly relevant. Torres never gives the impression that he had digested all the available modern evidence contained either in the *Teatro crítico* or in any other serious account of scientific investigations. He had the artist's impatience with finicking detail of procedure, and of prosaic illustration. He argued over sensational words and trends rather than over the minutiae of complex evidence, was too ready to adopt or manufacture solutions, explain problems away, to rely for his enlightened principles on jargon, and to base evidence on sensual appearances. The *Pepitoria crítica* had said that Torres did not answer his critics, he merely called them names. Perhaps it was hardly kind of the *Pepitoria* to mention publicly that the reason for the election of this Stargazer to the Salamancan Chair of Mathematics was the paucity of candidates. But Torres was not a man to be embarrassed into silence. The personal blows he delivered and those that he received are matched, quantitatively, with superb poetic justice. While, as the *Pepitoria* noticed,

Torres' protective method of defending his ideas – whether folk-based or semi-enlightened – was to attack personalities, he did not usually attack in isolation but followed some conspicuous lead which avoided the vital issue and influenced the development of irrelevant polemic. This was an activity not altogether disadvantageous to a period in need of mentally detective practice, one as yet unused to relying on laboratory test and apparatus and to exercising itself in marking time. Early in his polemical career he had been unconfessedly, possibly unknowingly, inspired in this way by one of Feijoo's first opponents on medical subjects, Dr Pedro Aquenza, to suggest that the Benedictine might be expected to employ his undoubted genius to best advantage by confining his scholarly activities to matters concerned with his religious profession. Torres would be unlikely to realize to what extent the personal form of his attack on Feijoo in the *Posdatas* of 1726 was due to the initiative of Aquenza's *Breves apuntamientos* . . . earlier in the same year.[17] When he followed the lead of other critics he obeyed an artistic instinct to develop their potentialities creatively. So his *Posdatas*, dedicated to Aquenza, go much further than the *Breves apuntamientos* by presenting Feijoo's secular *Teatro crítico* as work undertaken against the spirit of the Benedictine Rule. He protested that Feijoo had no right to dabble in Medicine without Papal authority; and that his duty was to restrict his activities to prayer, fasting, taking the discipline, and behaving in monkish character. Like other critics of Feijoo, Torres eventually was obliged for his own sake to desist not only from personal remarks against the Benedictine, but from academic criticism of his treatises. Feijoo was honoured by Fernando VI in 1748 with the title of Councillor for his 'profound, specialist learning and most useful words', as the reporting *Gaceta* put it.[18] It will be remembered that the royal decree forbidding further attacks on Feijoo's writings followed two years later.

As if, then, to compensate for caution exercised over Feijoo even in the 'twenties, Torres released his emotional inhibitions in increasingly personal attacks on Martínez, on Isla, offensively conspicuous in his trailing veils of pseudonymity, and on other critics masked and undisguised. Torres had his own forms of indirectness, his own schemes of assault by remote control. These were his fiction-frameworks which in polemic he decorated with a degree of loving care corresponding in intensity to the force of his feelings. One of his most effective Papers in this respect is his allegorical *Entierro* . . ., the 'burial' of Martínez's *Juicio Final* which was intended as a 'Final Judgment' against Torres' Astrology. Here, in the *Entierro* of 1727, characters, including Feijoo, gather for the funeral of Astrology, recently murdered by Final Judgment, only to be informed by Hippocrates that Astrology is still alive, and find themselves attending, instead, the funeral of the collapsed Final Judgment. Another piece of artistry is the satirical Correspondence with dead scholars in the *Correo del otro mundo*, where Torres speaks for and against himself with the help of such souls in the news as Aristotle, Hippocrates, and the original *Sarrabal*-almanac of Milan, which is neatly described as dead in its coffin among the worms without realizing the fact. For, we must remember, the eighteenth-century almanac *Sarrabal*, with which Torres competed, was merely a copy and continuation of the seventeenth-century publication. Torres' artistry even in scientific contexts was all too beguiling. Here his whimsical treatment of correspondence-terminology was enough to make his public forget that he was opposing serious scientific claims: 'I am in receipt of Your Mortality's letter';[19] '. . . requesting the discreet comments of Your Defunctness, my dear, dead Sir. . .'[20] Into the drama of such situations Torres could enter with the verve of the best satirists in Europe. As in some of his dramatic sketches, the scene reached its liveliest when he took part in person. So in the *Correo* he ironically puts his current almanac into the hands of the Sarrabal, and, with one eye on his admiring public, sketches the reactions of his master and rival in prediction, and repeats criticism from other correspondents about his own 'immodest pen', his general vanity, and his opinionated ignorance.[21]

The modern reader, like Torres' general public of the eighteenth century, is only too ready to ignore the rights and wrongs of polemical dialogue when the staging is so spectacular. But all this entrancing theatricality did not prevent Don Diego from venting his feelings in the form of realistic threats, or from extolling himself to the point of ludicrous unwisdom. Anger did not fetter his imagination. But it made him unsound in debate and reckless in self-defence. Consequently not only had he taunted Martínez about his luckless patients, but had threatened to compile and publish a list of those who had died.[22] He boasted unblushingly to one of his disguised critics, Fray Zutano, of being venerated as an astronomer throughout the length and breadth of France, Italy and Spain.[23] He threatened publicly to unmask an anonymous tormentor, whom he described as an expelled Jesuit, if that critic gave him further trouble.[24] No writer in that opinionated century had a stronger feeling for his own reputation than had the 'humble' Torres. On at least two occasions he even went so far as to describe himself, in his capacity as a cleric, as a 'Doctor of the Church' and apostle of the Lord,[25] and, oblivious of the fact that he had criticized the pen and person of a Father Master Professor of Theology, he claimed that anyone writing against Doctors of the Church and members of the University body were liable by Papal edict to excommunication. Certainly this assertion, with its ascription to himself of the title 'Doctor of the Church', he was later obliged to withdraw after friendly consultation, so he said, with the Inquisitor General, who doubtless was more amused than shocked. But even in his *Delación*,[26] or published retraction, Torres was able to bluff his way through a smiling explanation, with no suggestion of ignominy, that the phrase 'Doctor of the Church' and 'apostle of the Lord' constituted little matters of verbal inaccuracy, almost of misprint, occasioned by the confusing provocation of his libellous critics. Both attack and retraction are thoroughly characteristic. The ingenuity of Torres' unsoundness, indeed ignorance, set him marvellously conspicuous and apart in public favour and sometimes commended him even to General Inquisitors who, at least in the eighteenth century, were not always quite so narrow-minded or sadistic as foreigners liked to believe. In any case the influence of Torres' ideas both on halfway *ilustrados* and intelligent members of the general public must not be underestimated.

Several of the most topical subjects of international and national debate, classifiable perhaps as semi-scientific, were those concerning the transmutation of metals by a 'Philosophers' Stone', or the powder-quintessence ideally to be extracted from base metals; the explanation of magic, especially in miracles and witchcraft; the supposedly harmful effects of comets and eclipses. These border-territories of research were cultivated, as was natural for the times, by borderline thinking halfway between inherited philosophy and new pragmatism. More technical were discussions about the scientific evaluation of earthquakes, particularly in view of the earth tremors in Valencia in 1748 and the Lisbon earthquake of 1755; the nature of electricity, and other newly discovered scientific phenomena; the performance of newly invented instruments etc. One of the most obvious examples to select for observation of what was understood by 'evidence' is the international debate on the ancient Alchemist notion that under the right conditions, metals can be transmuted into gold. The most judicious of non-scientist scholars refrained from making dogmatic statements on the subject knowing that investigation into the properties of metals was as yet in its infancy. As late as 1775 Dr Johnson would still have liked, according to Boswell, to believe in the possibilities of transmutation, but was keeping an open mind.[27] When Feijoo, more scientifically alert than Johnson, was discussing the subject knowledgeably, from his reading of Boyle, in 1729, he sceptically disengaged attention from the over-general question: Can metals as imperfect entities be transmuted into the perfection of gold?, and directed it to the fundamental questions – Why should gold be considered the perfected metal? Why, if all other metals are imperfect forms of gold, should not the gold seams in mines show intermediate stages between perfection and imperfection? What in concrete fact is a metal, every individual metal? And what evidence, apart from unsubstantiated opinion, is there to suggest that all matter is a variety of the same basic element?[28] In other words, Feijoo's instinctive distrust penetrates to areas beneath mere speculation. With respective hope and scepticism these two scholars of enlightened associations waited for the results of modern laboratory experiment, and discounted 'evidence' of historical hearsay, before making definite pronouncements.

But the fact that international Johnsons would have liked to believe in the power of Alchemy caused a folk-majority, less cautious than they, to proceed to the 'evidence' of wishful thinking made unsubstantially constructive. Fray Luis de Flandes was one of those who, after taking into account the scientific difficulties involved in proving the case for transmutation, and after agreeing with sceptics about superstitious abuses, could confidently assert that transmutation is possible. His 'evidence' is of two kinds: historical, that is, from written reports of results by unspecified methods; and, more significantly, practical analogy. Regarding the first he is indebted to reports, traditionally accounted reliable, of transmutation, including transmutation effected by Ramon Lull, who achieved it allegedly by processes of divine intuition. Flandes put his faith in the impressive number of such reports. His example by analogy shows, however, that he was trying to think in an enlightened way by judging the general from the particular, without realizing that analogically two particulars in two different fields of research do not necessarily promote the same general principle. Transmutation, according to this, the analogical method, the *vulgo*-method of eighteenth-century thinking, must be possible if it can be shown to be possible in parallel cases of non-metal particulars. The parallel example chosen by Flandes to his own conclusive satisfaction is taken from medical sources:

> Cuando el manjar está recibido en el estómago y se convierte en sangre, . . . es verdadera transmutación.[29]

> (When food is received into the stomach and is converted into blood . . . that is true transmutation.)

Flandes' second volume, licensed also in 1742, repeats his objections to Feijoo's continued scepticism on the subject, declares that he can cite an additional forty authors to prove the existence of a 'disolviente universal' (universal dissolvent),[30] and concludes that if Feijoo cannot believe so great an array of witnesses he displays an obstinacy unworthy of a cleric.

The mind of 'know-all' Torres worked in similar fashion and

helped to establish him as arch-representative of middle-way enlightenment. Thus in *El ermitaño y Torres* he followed the new fashion for discrediting belief in the sensational and unsubstantiated accounts of supposedly successful Alchemists. Yet he clung to the traditional principle on which belief in transmutation was based and which Feijoo had tried to undermine: the assumption that all metals must surely be imperfect forms of gold. And while repeating current arguments that those professing to know the secret of the 'Stone' were suspiciously poor themselves, he saw any possible form of transmutation to be perfected in the future as the result of a manoeuvre properly engineered under star-influence by a process so far remaining secret.[31] For the most part, therefore, his distrust refers not to the Alchemists' ends but to their means and their persons.

The artist in Torres yearned for the suggestive invisibility of magic, an instinct he shared with many artists who inherited it from folk-needs. In their own way they used the conception of magic as a kind of Philosophers' quintessence for the ultimate transformation of base knowledge into a perfected reconciliation of natural and supernatural reality. Torres might be too modern to admit all stories of magical manifestations without examining them. But he retained belief in the possible existence of evil spirits exteriorized in witchcraft and, like Johnson, he referred to the 'evidence' of their confessions under torture, which Feijoo had denied to be evidence in such circumstances.[32] Torres' idea of evidence had little to do with the technical examination of a phenomenon. Not the testing but the appearance of reality was what determined his opinions. Like many more of his kind he could disagree, or rather echo those who disagreed, with Newton's theory about the elliptical form of earth, on the evidence of traditional doctrine alone. This was doctrine so cumulatively impressive that it could seem to constitute the force of authority,[33] just as the confessions of witches seemed cumulatively to constitute an authority difficult to ignore.

It is illuminating to compare the articles on earthquakes written by Torres and Feijoo in which the two authors wrote each in his characteristic brand of scientific intellectualism, and in which they diverge from each other in sharply contrasted technique.[34] Implicit in their divergent procedures are their divergent principles of intent. The well-read Feijoo, in academic detachment, economically came straight to three negative essentials: that the newly recognized phenomenon of electricity, already associated by scholars with earth-tremors, was a force as yet unknown and so as yet unassessable; that supposed signs of imminent earthquakes had been confused with effects; and that ancient theories about the whole subject, those of Pliny, for example, would have to be revised in the light of new, so far incompleted research. Torres, with the creative writer's desire to introduce ingenuity into statements of facts,[35] demonstrates his ingenuity on this occasion by comparing the earth and its inner activity to a human body with its pores, circulation of the blood, bowel-congestions and stomach-wind, its sulphur and inflamed spirits and its internal combustion. Electricity does not enter into the argument. His authorities are Pliny and Aristotle. It is an account rendered amusing and readable indeed, but Torres again, in his folk-instinct, avoids the fearsome emptiness of *Epoche* by analogical constructiveness.

Feijoo himself had said, argued Ignacio de Armesto y Ossorio, taking Feijoo out of context by generalizing from a relative statement, that our senses were provided to explain to us our natural environment. Which means, Armesto concludes incorrectly, that they can be relied on for purposes of factual proof. He had been speaking, as disapprovingly as Soto y Marne, of thermometers, reported by Feijoo to refute popular belief and, retorts Armesto, to mislead the 'filosófos más doctos' (most learned philosophers) that subterranean caves become warmer in winter and colder in summer.[36] Armesto had admitted absurdities in much astrological doctrine. Yet, like Soto y Marne, he could not bring himself to break with traditional belief in the dire effects on us of so abnormal a feature as an eclipse of the sun, whatever Feijoo might say about our everyday experience of 'eclipses' of light when we shut out the sun by going indoors to darkened rooms.[37] It would take time for the *vulgo* willingly to transfer reliance on sensual observation to reliance on dehumanized mechanism. Armesto's voice is one in an uneasy, even neurotic chorus, often dominated by Torres but including voices, such as Armesto's, more consistently honest than that of the Salamancan artist-astrologer.

To people trained to judge particulars from the standpoint of authoritatively established universals and to depend for evidence on externally sensory 'demonstration', the reasoning of inductive Science can be received by infiltration only: an unsatisfactory process, seemingly, to all concerned, but one to which society is accustomed. Change by permeation rather than by crisis or revolution certainly gives more time for misunderstanding, misinterpretations, and moral and intellectual confusion. Permeating unease, however, like mere fallowness, can help to divest the mind of stagnated certainties. Little by little, through the inconveniences of muddled debate, a drifting awareness of the meaning of 'inaccuracy', 'evidence', or 'proof', and of growing acceptance of uncertainties can merge, however uncomfortably, into healthy receptivity. This process, exemplified not only by the average educated Spaniard but by his European equivalent, is more noticeable in the Peninsula because history selects for scientific emphasis the work of individuals made universally conspicuous by its originality. And Spanish scientists of the eighteenth century could not compete for notice in that category. Their place in the history of inductive enlightenment is on the commoner level of debate, where scholars joined their general doubts to those of their neighbours. The new experiments in Applied Science and Medicine disturbed an international public, including the general public of countries becoming notable for 'enlightened' leadership. Beneath enlightened surfaces suspicion, often healthily critical, sometimes unhealthily ignorant, affected academic and non-academic attitudes and exposes realistically the period's mental state of natural complexity.

II Medicine

It will take France a long time, remarked the Journal of Trévoux in 1724,[38] to accustom herself to the idea of inoculation. British Presbyterians, observed Feijoo in 1733,[39] condemn the practice of vaccination on religious grounds. Boswell, in relation to Dr Johnson's background, mentions prolonged belief in England in the 'King's Evil'.[40] At the beginning of the century Pierre Le Brun had foreseen that medical dependence on Astrology would obstruct advancement.[41] In fact, reaction to outright *ilustrismo* denoted an international state of flux. While the Vatican could assert, as Riera reports from Piquer, that, in material cases of necessary doubt and sensory scepticism, one should follow the most probable explanation,[42] opinions could worriedly differ over the technical meaning of probability.

The greatest problem academically faced by eighteenth-century medical 'sceptics' concerned their abandoning certain basic Hippocratic tenets – that is, tenets ascribed to the Hippocratic School. Most of the converted experimentalists revered Hippocrates as an essential forerunner of modern medical Science, rather than as the inviolable authority sanctified for all time by misinterpretations of biblical injunctions to honour the physician.[43] As a forerunner, Hippocrates could continue to be esteemed for his practical observation of symptoms and development of disease about which he had probably learnt as much as his time and circumstances permitted. The moral standards he set concerning the practical care of patients and their individual dignity could not be faulted. Certain aspects of his medical teaching could still appear to be entirely relevant, especially, for instance, in relation to diseases which could be visibly assuaged by means of medicaments and external manual treatments, and for which dependence had to be left, for better or worse, to the internal workings of man's own nature. As yet few 'sceptics' would have been prepared to abandon the Hippocratic practice of bleeding and purging, and chiefly doubted in terms of the appropriate times and seasons for bleeding patients, and the amount of bloodletting to be prescribed for any one case. In general their arguments were concerned with where, how, and when Hippocratic methods should be brought up to date.

Academic and general opposition to the new medical 'sceptics' was partly concerned, then, with their presentation of Hippocrates as a mere forerunner of modern Medicine, and partly with their criticism of certain Hippocratic techniques such as the theoretically calculated timing of treatment to accord with pre-calculated crises of fever – stipulating the length of time during which fever should develop before any treatment is given – or the extent to which cures should be left to nature, and the disregard of possibilities of anatomical research. An obvious analogy with the effect of moon-movements on tides, weather, crops and animal behaviour had been a strongly predisposing factor in traditions of medical doctrine and was difficult to bypass. Much controversy, therefore, was conducted with circumstantial reasonableness over how much of established Hippocratic authority could be ignored, modified, or modernized without detriment to accepted Hippocratic standards.

This is well illustrated by numerous pamphlets which attempted to interpret Hippocrates in modern terms, by defending his divergence from the methodology of the Enlightened, or differentiating between what is permanently scientific in his scholarship and what necessitated alteration. The very titles of such articles – *Hipócrates defendido* (*Hippocrates Defended*)[44] (1711), *Hipócrates desagraviado* (*Hippocrates Indemnified*)[45] (1713), *Hipócrates vindicado* (*Hippocrates Vindicated*)[46] (1713), *Hipócrates aclarado* (*Hippocrates Clarified*)[47] (1716), *Hipócrates entendido* (*Hippocrates Comprehended*)[48] (1719) etc., indicate much of the nature of the epoch's medical disturbance among half-way *ilustrados*. A host of pamphlets on the same subjects, but with less obvious titles,[49] raised controversy to fever-level.

A good representative of unprejudiced but partially troubled *ilustrados* in a controversy over Hippocratic authority is Dr Miguel Marcelino Boix, Professor of Medicine in the University of Alcalá and Founding Member of the Sociedad Regia de Sevilla, whose *Hipócrates defendido* (1711), written apologetically in Spanish though lavish in Latin quotation, is as good an example as any of thoughtful, and so academically significant concern. In 1722 he was to act as one of the sponsors of Martínez's *Medicina scéptica*, a work which seemed to him to lead in a right direction, apart, he says, discreetly, from certain points which he had communicated to Martínez privately.[50] The exception, however private, is illuminatingly typical of his caution. Boix could sympathize with enlightened empiricism; could approve of Bacon, Boyle, Baglivi, Sydenham, Etmüller, Harvey, and their compeers; appreciate the force of their determination to suspend judgment where evidence was unobtainable or doubtful. But typically he represents European mid-way thinking, and so personal and national discomfiture, by needing to believe and prove that any such acceptable development of scientific thinking was a development in line with Hippocratic guidance. In other words, he was hampered by the common need, analagous socially to religious convictions, to rest from nagging doubts in the assurance of a timeless, unassailable authority.

An obsessive need for established authority is therefore the hallmark of earnest semi-*ilustrados* who, from a twentieth-century standpoint, wasted their energies in defences, interpretations, and re-interpretations of the Hippocratic system. Their arguments tend to develop incidental points of controversy by theoretic generalizations (especially, for example, on the subject of man's 'nature', and on the reduction of the infinite variety of diseases to established categories amenable to 'nature's' cures), are affected incidentally by Scholastic treatment of Hippocratic terminology, normally evince a certain horror at the idea of carrying unlimited anatomical experiment and the development of surgery to what seemed unhuman extremes, subordinate the analysis of cause to that of effect, and, like all defences and criticisms of Hippocrates, become involved in discussions over the validity of Hippocratic texts. Their arguments, needless to say, miss the larger points of enlightened medical research. Matter loses to method. Boix's technique, natural to most academic scientists of the time, was to analyse meticulously the Latin terminology of Hippocratic aphorisms in supposed Hippocratic texts, and so to be limited by the presuppositions of Hippocratic signposts. Only the untypical few in Europe could understand that extensive areas of medical ground must be thought of as uncharted. Consequently the points discussed by Boix have little to do with implications of new discoveries, and deal rather with the incidental relationship of Hippocratic and rationalist practice, particularly with regard to times of bleeding and purging. For since the different stages of 'cures' by such generalized treatments as bleeding and purging could still not be established in practical detail, the use of delaying tactics in the 'natural' application of bleeding and purging, to conform with 'natural' astrological conditions, still made 'natural' sense.

Sydenham, for instance, could be regarded as 'uno de los mayores prácticos que tiene la Facultad Médica' (one of the greatest experts within the Medical Faculty).[51] The modern 'scépticos' are to be applauded for their discoveries. Indeed, says Boix, St Augustine had approved of 'sceptics' in non-religious thinking, and Hippocrates himself 'nos amonesta que suspendamos el juicio' (warns us to suspend our judgment).[52] Which means, he later concludes, 'que el hombre no tiene criterio, potencia, o facultad en su entendimiento, para poder conocer la esencia de la entidad más mínima, que contiene este universo' (that man has not in his understanding the criterion, power or faculty to be able to identify the essence of the most minute entity contained in this universe). The scientist, therefore, thinks Boix, cannot make discoveries in an uncharted void. Discovery must be situated within known, that is Hippocratic, guidelines.[53] Consequently, we gather that what was lacking for the total conversion of the semi-*ilustrado* was prescribed definition, a revised, new system to unite the new author authoritatively with the old. The lack of an explicit systematization would remain a major source of confusion. To date, only Hippocrates had supplied what seemed to be that vital need:

> . . . el sistema que menos afán trae en la curación de sus enfermos, y el que menos remedios les aplica; y sobretodo, el que más deja obrar la Naturaleza . . .[54]
>
> (. . . the system causing least trouble in the healing of his patients, and that which involves the fewest treatments; and above all that which best leaves the work to Nature . . .)

Hippocrates had constructed a medical map, which, if not appropriately extended, must at least be replaced, it was commonly felt, with a positive equivalent.

'Suspension of judgment', the watchword issuing proudly from the mouths of professedly advanced scholars, could so easily be misunderstood by them. Boix is typical of those who use it naively. Neither he nor any other semi-*ilustrado* could assess its ultimate significance. So his 'suspension of judgment' was partial. Perhaps he assumed, in his urgent support of Hippocrates, that the latter was 'suspending judgment' by leaving cures as far as possible to 'nature': a method safer, indeed, than the application of many ancient remedies, and much lauded by cautious semi-*ilustrados*. Boix himself had reasonably refused to bleed patients affected by an epidemic of smallpox, though bleeding was a traditional remedy for that complaint, and relied on a mild prescription of barley water and 'nature''s own processes:

> . . . si se morían algunos, más era por culpa suya y mal asistidos, que por la misma enfermedad.[55]
>
> (. . . if some died it was rather through their own fault or because they were badly served [by others] than because of the nature of the complaint.)

It did not, however, occur to him to think that Hippocrates' conception of 'nature', and so his own, might not remain a mystery for ever, and that its as yet unknown identity and processes could possibly be determined by as yet unknown means and openended research. Nor, on the controversial subject of bleeding and purging and other traditional treatments expected by patients as practical, and so immediately sedative, remedies, could he conceive of that subject as largely irrelevant, in emphasis and incidental detail, to unpredjudiced discussion of Anatomists' new findings about disease and what clinically might contribute to new methods of diagnosis and treatments.

Every now and again Boix, like other semi-*ilustrados*, impresses us with his desire to see more than one point of view. What, he asks sensibly, do we mean by the scientific word 'demonstration'? Philosophers, he observes, do not 'demonstrate' the meaning of, for example, a compass, when they explore theoretically the probable reason for its operation. 'Demonstration' in this case is the sailors' practical experience of how the compass works for them. At other times he clings to unproved tradition: to the need for astrological calculation in timing medical treatments. 'Care must . . . be taken', Hippocrates had pronounced in this respect, 'at the rising of certain stars, particularly the Dog Star and Arcturus. Similarly, discretion must be exercised at the setting of the Pleiads. It is at such times that the crisis is reached in the course of diseases; some prove fatal and some are cured, but all show some kind of change and enter a new phase.'[56] And the natural, folk-force behind astrological reasoning conditions Boix's acceptance of astrological importance. The medical practitioner, then, he says, in addition to knowing Latin, Greek and the new Natural Philosophy (that is, not Natural Philosophy taught in traditionally academic schools as a theoretic discipline) must know 'su poco de Astrología'(his bit of Astrology).[57] Ironically, on the negative side, Boix could dispute the essential value of extending the practice of clinical Anatomy and of giving too much importance to studying the circulation of the blood. Such researches take Medicine away from medical practice, he, in his eighteenth-century circumstances, not unnaturally thought. Hippocrates had been no Anatomist, and had been ignorant of the circulation of the blood. Yet

> . . .en medio que ignoró todo esto, fue, y es el mayor médico que han conocido todos los siglos, luego para ser buen médico no se requiere tanta Anatomia como quiere Bartholino . . .[58]
>
> (. . . while understanding none of this, he was, and is the greatest medical doctor known throughout the centuries. Consequently in order to be a good doctor not as much Anatomy is required as Bartholino wishes.)

So you cannot, he concludes, be a good doctor and a good Anatomist.

Which means that the semi-*ilustrado* in general could not yet appreciate the newly emerging distinction between the specialist and the general practitioner. From the Hippocratic standpoint the noble and moral standards of Medicine were based on general practice, on homely rather than clinical relations between doctor and patient. Even so, the very fact that Boix could face the desirability, if not the necessity, of some forms of medical research that might extend Hippocratic foundations, could be as wrongly disturbing to some as it was rightly disturbing to others. As the poet Antonio Zamora pipes up unexpectedly in a sponsoring 'Romance poco-serio' (Not very serious Ballad) to Boix's treatise:

> Vuestro libro (como todos)
> tendrá sus apasionados;
> y la prueba de que es bueno
> será, tenerle por malo . . .[59]
>
> (Your book [like all books]
> will have its fervent supporters;
> and the proof that it is a good book
> will be that it is taken to be a bad one . . .)

All heartfelt arguments among reasonable scholars over the permanent dictatorship of Hippocrates in the standardization of guidelines revolve round the same theoretic irrelevances rendered incidentally more irrelevant for modern purposes by reliance on Hippocratic terminology. Boix's *Hipócrates aclarado* of 1716, reacting to criticism of *Hipócrates defendido*, pursues the interpretation, and what he considers misinterpretation, of Hippocrates' aphorisms, for instance, in reference to the obsessive subject of 'natural' timing for *sangrías* (bloodletting). His three aims of clarification again concern theoretical interpretation of a Hippocrates 'mal entendido' (ill understood).[60] Álvarez del Corral, in his *Hipócrates vindicado*, plunges into the controversy over Hippocratic texts and authorship, and urgently stresses the importance of acknowledging that Hippocratic Medicine has modern applications.[61] Antonio Díaz del Castillo, in his *Hipócrates desagraviado*, finding that Boix has not spoken positively enough in his support of Hippocrates, also engages in terminological discussion over interpretations of 'nature', 'scepticism', disease, etc.[62] And later, in *Hipócrates entendido* of 1719, replying to Boix's *Hipócrates aclarado*, he is still concerned with problems over Hippocratic-Galenic authorship discussed by Boix; still requires more 'positive' decision about the meaning of the word 'nature',[63] and is worried by Boix's objection to Galen's ideas which seem to contradict modern findings on the circulation of the blood. While Lessaca in *Formas ilustradas* of 1717 had found that those who refuse to accept the idea of the circulation of the blood are not unreasonable,[64] he himself, had he been given more theoretical reasoning and so more philosophical support for the idea, would seem not unwilling to accept it. Bonamich, in *Duelos médicos* of 1741, can praise Sydenham, but expresses unwillingness to desert speculative study. All are preoccupied with a need for the definitive authority of a newly constructed and all embracing Philosophy of Medicine, and tend to confine their practical attention to interpretations of 'Hippocratic' teaching on 'nature', and on the treatments limited, by non-clinical research, to such 'natural' measures as bleeding and purging.

Nowadays so many of their arguments must seem ironically foolish. But at least some of these semi-*ilustrados*, including Boix, were attempting to accept baffling facts of current discovery by interpreting them in familiarly philosophical terms and reconciling them, as they thought, with Hippocratic predisposition. Bonamich expresses the situation as follows:

> No quisiera incurrir en la nota de adherido a las modernas doctrinas, ni aficionado nimiamente a las antiguas opiniones, porque de ambas tomo lo más conforme, y dejo lo que no me parece consonante a razón. Ni desprecio absolutamente a Galeno, ni totalmente apruebo a Paracelso . . .[65]
>
> (I shouldn't want to commit myself to an outright adherence to the modern doctrines, nor to excessive enthusiasm for ancient opinions, because I accept what is most relevant from both, and I reject what does not seem to me to be reasonable. I neither absolutely despise Galen, nor do I totally approve of Paracelsus.)

These were serious scholars who seriously tried, within Hippocratically established limits, to practise academic openmindedness. Not so was the popular 'Scientist' Torres Villarroel, whose 'openmindedness' depended on his mood and circumstances, and whose semi-*ilustrismo*, therefore, was more emotional than academic, more artistic than scientific, and more influential publicly than that of the aforementioned academicians. Yet again he emerges as the most generally conspicuous, because most popularly comprehensible, of the half-way *ilustrados*, and because his conception of 'demonstration' and 'evidence' in Medicine, as in his scientific contexts, is illustrated with artistic provocation by folk-examples and analogies of an obvious, concrete nature.

Torres, the Mathematician, who seemed to regard Medicine as his second field of specialization, had been pronouncing on the subject almost from the start of his writing career. Sometimes, as in parts of the *Viaje fantástico* of 1724, his vademecum of the physical universe, he merely simplified and popularized established medical text-books, Gilabert's *Escrutinio físico-médico-anatómico*,[66] for example, by briefly describing, in non-technical language, the forms and functions of different parts of the body and the symptoms of diseases to which each part is vulnerable. At times he was discursively critical of current medical knowledge, as in *La barca de Aqueronte*. When he wrote a Paper on the supposed miracle, reported in 1747, of a sweating corpse in a mortuary, he dismissed the miraculous element with a scepticism worthy of Feijoo,[67] and this despite his artistic love of marvels. Moreover, using similar examples to those given by Feijoo of corpse-movements and blood-liquidity, he repeated injunctions, long stressed by the Benedictine, about the dangers of hasty interment.[68] What he did not seem to consider in the case of the sweating corpse were the difficulties, enumerated by Feijoo, of knowing just at which moment a man actually dies. Torres seldom encompassed any one problem fully. At his medical best, he significantly voiced the general opinions of his progressive contemporaries in his general disillusionment with medical practice to date. He joined in criticism, most notably publicized by Feijoo in 1726, of over-prescribing, over-bleeding, and over-purging, through the belief that purgatives removed selectively only what doctors wanted them to remove;[69] and in advocating, at least for minor ailments, the gentler nature-treatments then becoming fashionable, as a reaction to the use of drastic drugs and surgery: the herbal, thermal and hydropathic cures, cleanliness, fresh air, and exercise.[70] That he, Torres, was ready to bleed and purge with the best, and that some of his recommended draughts and ointments composed of animal, even human urine, or white excretions of hen cooked in wine,[71] now sound barbarous, was the fault of his century. Urine, to judge by its inclusion in so many folk-remedies of the past, must have been effective in certain cases at least for immediate relief. The eighteenth century was as yet unfamiliar with the full danger of toxic infection. As for bleeding and purging, no substitute for dispelling poisonous swelling had yet been discovered. It was true that the instruction to leave treatments to nature, as do the beasts, had a logic reasonably based on observation, especially since nature, in the common experience of those times, regularly indicated the need for induced vomiting and judicious purging to relieve the results of over-indulgence in food and drink. If cats eat grass to induce vomiting at certain times and seasons, Torres might reasonably have specified, there must be some natural reason for inducing vomiting regularly in human beings. The trouble is, said Feijoo, who thought on deeper levels, what exactly do we mean by 'nature'?[72]

The relation between bloodletting and Astrology had raised one of the most serious of *vulgo*-problems. The Astrologer Torres, whose reputation academically and publicly, partly rested on the confidence of his assertions concerning the importance of star-influence in human affairs, based his authority on well established, international tradition. Scientifically viewed, he probably was at his blindest, or most conveniently self-deceptive, in this very area in which he was most influential. As a popular Spanish equivalent to Old Moore[73] on the one hand, and on the other, as a Professor of Mathematics who professed to move with the times in pursuit of 'demonstrable' evidence, he was obliged to try to wrest astrological significance from astronomical facts, to the disgust of a few and delight of the many. Torres, Feijoo, Martínez, Aquenza, Isla, Ribera, Mañer . . ., to mention outstanding names, engaged vehemently in polemic on the subject. Their arguments concerned the ancient medical practice of consulting the influential state of the heavens when administering treatments such as bloodletting and purging; when calculating crises in nervous disorders such as melancholia, nervous depression, or the hysteria traditionally associated with a full moon; and when plotting the stages of fevers. Much incidental discussion developed over terminology, especially over the word 'influence', or 'effect'. In 1728 Salvador Joseph Mañer in his *Repaso general de todos los escritos del Bachiller Don Diego de Torres* . . . made a point of detailing Torres' self-contradictions in his use of the words 'truth', 'evidence' and 'demonstrable' especially relating to Astrology and Medicine. 'Torres always talks in absolutes', is one of his comments which rightly testifies to Torres' *vulgo*-values in Medicine and elsewhere.[74] Nor do Torres' expedient manoeuvres to relate Medicine to astrological Astronomy escape his indignant notice. Tradition, as we have seen, and as followed by Torres, allocated certain times and seasons for bleeding when star-positions were propitious to that exercise. Mañer's attitude here is that of Martínez, who had insisted that the time for treatment of any kind, including bleeding, must be dictated not by the stars but by the patient's immediate needs.[75] Bloodletting in any case was a dubious, or dangerous, remedy. So whatever the truth about the possibility of star-influence on bloodletting, concludes Mañer irritably, 'wouldn't it be better to be killed slowly by straightforward bleedings than to be smashed at one stroke into a sky-timed omelette?'[76] Nevertheless, in this respect as in so many others, it was Torres who represented the general medical understanding of man's relationship with his natural environment, and the astrological factor was one of the most difficult to reform: a point illustrated urgently by many of Torres' contemporaries.[77]

Torres' medical authorities had been challenged as early as 1726 when Isla wondered in print who, medically speaking, Torres thought he was to pontificate; when the Jesuit spelt out the meaning of medical scepticism as openmindedness; and when, apparently with some justification, he hinted at the inferiority of Torres' medical qualifications from so sub-standard an institution as was then the College in

Ávila.[78] From the beginning, therefore, the medical utterances of Doctor Torres tend to be as defensively combative as his pronouncements on Astrology, and usually indicate that understandably startled reaction to threats against the security of tradition experienced by any herd-mind in any country. Of current medical topics drawn to his attention, one was the new importance given to Anatomy and dissection of corpses: a developing interest which was to become the basis for modern investigations in European Schools of Medicine as influenced by, for instance, the University of Edinburgh. Torres approached this subject in the *Viaje fantástico*, the *Correo del otro mundo* and the *Desahuciados del mundo y de la Gloria* with near contempt, probably because dissection in Spain had normally been conducted by technical assistants to University Professors, as a form of manual labour, to give merely a general idea of the structure of the human body. It would seem to Torres to be a purely mechanical exercise: an operation mounted for purposes of exemplifying rather than as a process of discovery. He liked to think of himself as an empiricist in living relation to living activity, and believed it more 'demonstrable' to observe the living than the dead. It was impossible, he supposed, to deduce much from dead eyes, for example.[79] And indeed he was right to assume that impossibility while the dissector confined his attention to cataloguing qualities already known, and while investigators did not apply new knowledge in Chemistry to medical analysis. Torres helps to prove that barriers erected by philosophers of Science restricted scientific practice through departmentalization.

Among other current trends in Medicine in which Torres expressed interest was the emphasis given to the 'egg' discovery, or microscopic revelation that plant and animal life is composed of living cell-eggs. This, in the *Viaje fantástico*, he seemed to take for granted. He had, too, a limited understanding of infection, rather less limited than that of the average physician of his time, and, in the *Vida natural y católica*, the *Uso y provecho de las aguas de Támames . . .*, and other writings, he related infection to dirt, bad smells, lack of ventilation and infested water supplies. His knowledge might not extend to ideas about blood-groups, blood-transfusion and the benefits of inoculation against smallpox, discussed with great interest by Feijoo. But he could relate certain physiological causes to certain effects with a fair degree of accuracy, though sometimes by forcing enlightened medical knowledge and folklore into an uneasy alliance. A paper he submitted for a competition organized by the Royal Medical Association of Our Lady of Hope in 1750 may serve as an illustration. The question set by the Association ran as follows: 'In view of the fact that the normal habitat of worms found in human bodies is in the intestines, why do such worms produce rashes in the patient's nostrils?'.[80] Torres' answer for the most part was laudably technical. He derived the worms from microbe-infested food consumed by the patient; he related the intestines to the rest of the body by what we should now call the bile duct, and by the air passages; and he accounted for the upward movement of 'eggs', half or fully developed into microbes, as the result of mechanical propulsion through waves of respiration: a theory not too far removed, we understand, from the one accepted nowadays. But, for good measure, he went on to prescribe for the condition, and, in doing so, retreated into the past. His suggestion was to bathe the outer part of the affected areas, first with warm water, and afterwards with cock's blood, through which, he assured us, the worms would thrust their pin-point heads allowing us conveniently to shave them off. He did not win the prize. Perhaps the Association took itself too seriously to accept the typically jocular tone of Torres' typically jocular preliminaries in setting forth his thesis. Moreover, by 1750, Torres' scientific views had become, for all their popular raciness, academically discredited. Perhaps the Association disliked the manner in which he finally turned the tables by putting a question of his own, one, in fact, more pertinent than the original. Why is it, then, he countered, that other bowel complaints, colic, hernia, or any inflammation, do not also produce rashes in the nostrils?

In general, Torres' need, which was the need of all but the most advanced of his contemporaries, of a reassuring new system to replace the toppling systems of the past, applied to Medicine as surely as to Astronomy. An awareness that too many people died, or chronically ailed, from too many unknown causes, resulted in emotional cries for positive guidance. 'From Martínez's title "Sceptical Medicine", I expected a new system', Torres protested significantly in the *Ermitaño y Torres*, 'and all I found was the author's insistence on medical uncertainty'.[81] For he could not understand that applied Cartesian doubt, and Baconian, Newtonian and Feijoost openmindedness, was a first stage in medical rehabilitation. In the disappointing circumstances Torres and his kind tended to reason along the only safe line they knew, seeking reassuring constructiveness in inverse ratio to the negative degrees of Martínez's scepticism. From the relatively neutral ground of the *Ermitaño* Torres sought shelter, in the *Vida natural y católica*, of 1730, in current suggestions about natural living, and came to concrete conclusions about the moral causes of disease in the *Desahuciados del mundo y de la Gloria* of 1736. His nature-stand was made on the incompetence of doctors, a subject on which, in the *Vida natural y católica*, he expressed himself libellously by saying that he knew of 'no doctor who lived as God commanded and as nature taught', that the doctors' commerce is 'thieving and lying', – and had the passage in the 1743 edition cut by the censor.[82] Moving to the moral frailty of patients, he passed to positive codes of advice on nature's cures, and did not pause long enough with the new rationalists to distinguish clearly between codes of morals and codes of health. The very fact that self-indulgence has always been recognized as a cause of many diseases provided him with a form of 'demonstrable' evidence. Since, for example, an abuse of eating and drinking tends to produce apoplexy, or, we now say, high blood pressure, and since sexual promiscuity is liable to cause venereal disease, Torres felt encouraged, as ancient systems collapsed around his mind, to erect an over-facile system of medical cause and effect based on moral case-histories.

This is the general theme of the *Desahuciados* in which, determinedly passing from one fetid ward to another, Torres supposes that his closeness of examination, especially of the more repulsive aspects of sickness – patients' urine, stools, blood, furred tongues, vomit – as well as of pulse-rates, behaviour and previous modes of life, can of itself be equated with medical diagnosis. Feijoo, indeed, had advocated such careful examination considerably earlier, in 1726, not, however, as an end in itself, but as a preliminary to developments in analysis as yet unknown. So, 'that is not the point', he might have said to Torres.[83] The unknown quantity in medical research was more concerned with discovering, instrumentally, the chemical nature of bacteria, and the chemical means of creating resistance: an end towards which only an amoral scepticism could lead. Torres' method of procedure by obvious analogies between disease and human perversions was a natural one, but it was more literary than scientific, and it kept him tied, unsuitably on his own showing, to the pulpit.

It should be added in this medical connection that the censors of 1743 must have felt that Torres in the 1730 version of *Vida natural* . . . had been too explicit and too ambiguous in his discussions on the act of sexual intercourse: explicit, for the period, in the fulness of its detail; ambiguous, in that, while stressing the need, for purposes of male health, to use the sexual act, he failed to specify explicitly that it should be used only by married men with their wives.[84] The section was removed, with other sections on dubious morals in which, for example, Torres permitted a man to kill for the sake of his honour, and in which he related pride too closely to magnanimity. By 1743 Bourbon ideas about the social iniquity of duelling were beginning to take shape, and by 1743 Torres had learned from exile to view the sword with wariness. In any case, his opinions, whether on Medicine or Ethics, were not infrequently polluted by emotional imaginings.

'Evidence' among intelligent, semi-*ilustrados*, then, could mean systematized classification of the facts of exterior observation – furred tongue, vomit, abnormal excreta, sores, venereal infections etc., – that were partially or wholly unrelated to observations, with new instruments, and by internal analysis, of Anatomists and Chemists whose research took the form of unprejudiced and indeterminate detection. Such 'evidence' by classification alone related reassuringly, in whole, or in part, to the necessarily limited explanations of inherited medical dogma. In Medicine, as elsewhere, a natural reason for alarm was a vaguely recognizable threat of endlessly discoverable areas of inexplicability.

NOTES

1 See Russell P. Sebold's Introduction to Torres' *Visiones y visitas*, Clásicos Castellanos 161 (Madrid: Espasa Calpe, 1966), pp. xliv, *et passim*, and McClelland, *Diego de Torres Villarroel*, p. 140. A Practical argument about the meaning of evidence was broached by Isla in *Cartas de Juan de la Encina, contra un libro que escribió Don José de Carmona, cirujano de . . . Segovia. Carta Primera*. BAE, XV, pp. 404 ff.

2 See pp. 70ff. and note 40 below.

3 '. . . es hombre que de todo sabe': Salvador Joseph Mañer, *Repaso general de todos los escritos del Bach. Don Diego de Torres*, (Madrid, 1728), p. 6.

4 *Cartilla rústica*, 1727. See *Libros en que están retratados diferentes cuadernos físicos . . .*, *Obras completas* (Salamanca, 1752), 14 vols. Vol. VI, pp. 126 ff.

5 *Desprecios prácticos del Piscator de Salamanca . . .* (Madrid and Sevilla, 1725), pp. 8 ff.

6 *Correo del otro mundo al Gran Piscator de Salamanca*, Salamanca, n.d. (Dedication, 1725), pp. 28-29.

7 *Posdatas a Martínez*. The Dedication to the pamphlet we have used is dated Salamanca, 1726. See pp. 19 ff; p. 38. See also *Obras*, Vol. X.

8 See, for example, J. Rodríguez Espartero, *Reparos de encuentro . . .* (Madrid and Sevilla. Undated [1727]: an attack on Torres' *Visiones . . .* and chiefly notable for its triviality.

9 *Consejos amigables a Don Diego de Torres . . .* (Madrid and Sevilla, undated), pp. 3 ff.

10 Rodríguez Espartero, *op.cit.*, Respuesta 1a.

11 *Repaso general de todos los escritos de . . . Torres* (Madrid, 1728), Point 17.

12 *Diario de los literatos de España . . .* (Madrid, 1737), Vol. II, p. 299.

13 *Resurrección del Diario de Madrid, o Nuevo Cordón crítico general de España* (Madrid, 1748). See Prólogo, and pp. 14, 41 ff; 46 ff. *et passim.*

14 'Pedro Fernández' (Isla), *Glosas interlineales . . . a las Posdatas de Torres . . .* in *Colección de papeles crítico-apologéticos* (Madrid, 1788), Part I, pp. 71 ff.

15 *Juicio final de la Astrología . . .* (Madrid and Sevilla, undated).

16 *Pepitoria crítica* (Sevilla. Undated). See p. 21.

17 *Breves apuntamientos . . .* (Madrid, 1726). See pp. 8 ff. Aquenza's pamphlet was announced in the *Gaceta*: 22.X. 1726. Torres' *Posdatas* was announced over a month later in the *Gaceta*: 3. XII. 1726.

18 *Gaceta*, 19.X. 1748.

19 *Correo del otro mundo . . .*, *ed.cit.*, p. 22.

20 *Op.cit.*, p. 47.
21 *Op.cit.*, p. 86 and pp. 60 ff.
22 *Entierro del Juicio Final. Obras, ed.cit.*, Vol. X, pp. 140 ff.
23 *Insinuaciones a cierto apóstata satírico* . . . (1738). *Obras, ed.cit.*, Vol. X.
24 *Copia de una carta* . . . (1748). *Obras, ed.cit.*, Vol. X.
25 *Insinuaciones* . . . *Obras, ed.cit.*, Vol. X, pp. 252 ff.
26 *Delación* . . ., *Obras, ed.cit.*, Vol X, pp. 267-68.
27 Boswell, *Life of Samuel Johnson, ed.cit.*, Vol. I, p. 566.
28 See McClelland, *Benito Jerónimo Feijoo*, pp. 75-76.
29 Flandes, *El antiguo académico, contra el moderno scéptico, ed.cit.*, Vol. I, p. 214.
30 *Op.cit.*, Vol. II, p. 180.
31 See, for instance, Torres' *Viaje fantástico, ed.cit.*, pp. 18 ff. and *El ermitaño y Torres, ed.cit.*, pp. 50 ff.
32 See Torres, *Viaje fantástico, ed.cit.*, pp. 237 ff; Boswell, *Life of Samuel Johnson, ed.cit.*, Vol. I, pp. 429-30; Boswell, *Journal of a Tour to the Hebrides with Samuel Johnson*, Everyman, No. 387 (London, 1935), p. 29; McClelland, *Benito Jerónimo Feijoo*, p. 85.
33 See *Prevenciones que le parecen precisas a* . . . *antes de entrar a la narración de las observaciones con que se intenta persuadir, que es elipsoides la figura de la tierra* . . ., *Obras, ed.cit.*, Vol. IV.
34 Torres, *Tratados físicos y médicos. De los temblores* . . . *de la tierra*. See *Obras, ed.cit.*, Vol. V; Feijoo, *Cartas eruditas, ed.cit.*, Vol. V, Nos. 13-14.
35 Torres, *op.cit.*, p. 47.
36 Ignacio de Armesto y Ossorio, *Teatro anti-crítico universal* . . . Undated (Dedications, 1735-37), Book III. See pp. 245 ff.
37 *Op.cit.* See Book II. Feijoo had dealt with the subject of thermometer-evidence in caves in 'Del antiperístasis': *Teatro crítico universal, ed.cit.*, Vol. II, Discurso 13; and eclipses in 'Eclipses', *op.cit.*, Vol. I, Discurso 9. For Soto y Marne's views on these subjects, see his *Reflexiones crítico-apologéticas sobre las obras del RRP Maestro* . . . *Feijoo* (Salamanca, undated [1748-49]), Vol. I, pp. 172 ff; Vol. II, Reflexión 23.
38 See *Mémoires pour l'Histoire des Sciences et des Beaux Arts* (Trévoux, June 1724).
39 *Teatro crítico universal, ed.cit.*, 1733. Vol. V, Discurso XI 'El gran magisterio de la naturaleza'. See Section 14.
40 *The Life of Samuel Johnson, ed.cit.*, Vol. I, p. 17.
41 Pierre le Brun, *Histoire critique des pratiques supersticieuses qui ont séduit les peuples et embarrassé les savants* (Rouen, 1702). It is incorporated in *Superstitions anciennes et modernes; préjugés vulgaires qui ont induit les peuples à des usages et à des pratiques contraires à la réligion* (Amsterdam, 1733), 2 vols. See Vol. I, Book III.
42 See Juan Riera, *Medicina y ciencia en la España ilustrada, Epistolario y Documentos*, I, p. 105. Regarding Piquer's conception of modern Physics and the limits to rationalistic thinking in such advanced Spanish *ilustrados* as Mayáns, see Vicent Peset, *Gregori Mayans i la cultura de la Il·lustració*, pp. 270 ff, *et passim*. A useful list of medical and scientific publications between 1712 and 1750 is given by Iris M. Zavala, *Clandestinidad y libertinaje erudito en los albores del siglo XVIII* (Barcelona: Editorial Ariel, 1978), pp. 419-24.
43 See p. 33 above.
44 Miguel Marcelino Boix, *Hipócrates defendido de las imposturas y calumnias, que algunos médicos, poco cautos, le imputan* . . . (Madrid, 1711).
45 Antonio Díaz del Castillo, *Hipócrates desagraviado, de las ofensas por Hipócrates defendido* . . . (Alcalá, 1713).

46 Antonio Álvarez del Corral, *Hipócrates vindicado, y reflexiones médicas, sobre el Hipócrates defendido* (Madrid, 1713).

47 M.M. Boix y Moliner, *Hipócrates aclarado, y sistema de Galeno impugnado, por estar fundado sobre dos Aforismos de Hipócrates no bien entendidos . . .* (Madrid, 1716).

48 A. Díaz del Castillo, *Hipócrates entendido, a beneficio de la doctrina de Galeno, su fiel intérprete . . .* (Madrid, 1719).

49 See, for example, Juan Martínez de Lessaca, *Formas ilustradas a la luz de la razón . . .* (Madrid, 1717); Narciso Bonamich, *Duelos médicos . . .* (Madrid, 1741); Juan de Adeva Pacheco, *Verdadera medicina . . .* (Madrid, 1754), etc., etc.

50 See *Medicina scéptica, ed.cit.*, Vol. I, (Second) Approbation.

51 *Hipócrates defendido, ed.cit.*, p. 73.

52 *Op.cit.*, p. 206.

53 *Op.cit.*, pp. 248, 276.

54 *Op.cit.*, Dedication.

55 *Op.cit.*, p. 133.

56 See *Hippocratic Writings* (Penguin Classics), p. 159.

57 Boix, *op.cit.*, p. 281.

58 *Op.cit.*, p. 285.

59 *Op.cit.*, see preliminary section of Approbations.

60 *Hipócrates aclarado . . ., ed.cit.*, p. 31.

61 *Hipócrates vindicado . . ., ed.cit.* See Prólogo, *et passim.*

62 *Hipócrates desagraviado . . ., ed.cit.* See, for example, Discurso II 'En que se trata qué cosa sea la naturaleza', etc. On scepticism see, for example, p. 483; *et passim.*

63 *Hipócrates entendido, ed.cit.* See, for example, Prólogo; pp. 287 ff; *et passim.*

64 *Formas ilustradas . . ., ed.cit.* See Chapter III 'Del movimiento circular de la sangre'.

65 *Duelos médicos . . ., ed.cit.*, p. 193.

66 See Torres, *El ermitaño y Torres. Obras, ed.cit.*, Vol. VI, p. 18.

67 See *Desengaños razonables . . . Obras, ed.cit.*, Vol. IV.

68 Feijoo, *Teatro crítico universal, ed.cit.*, Vol. V (1733) Discurso VI.

69 *Desengaños razonables, ed.cit.*, Vol. IV; Feijoo, *op.cit.*, Vol. I, Discursos V, VI.

70 See *Uso y provechos de las aguas de Támames* (1744). *Obras, ed.cit.*, Vol. IV, and *El doctor a pie . . .* (Salamanca and Sevilla, 1731), etc.

71 *Vida natural y católica . . ., Obras, ed.cit.*, Vol. IV, p. 65.

72 *Cartas eruditas, ed.cit.*, Vol. III, no. 6.

73 See McClelland, *Diego de Torres Villarroel*, pp. 133 ff.

74 *Repaso general . . ., ed.cit.* See pp. 66 ff.

75 See Martínez's sponsoring *Carta defensiva . . .* in Feijoo's second edition of the *Teatro crítico universal*, p. 9.

76 *Repaso general . . ., ed.cit.*, p. 50.

77 On the importance of Astrological relations to Medical treatment, see also, for example, the typical, half-way *ilustrado* Pedro Enguera, *El Gran Gottardo*, 1723; Juan de Fuentes in *El Piscator de la Mancha. Diario médico*, 1731; Gómez Arias, *El decreto de Minerva*, 1749, etc., etc.

78 *Glosas interlineales . . . a las Posdatas de Torres* in *Colección de papeles crítico-apologéticos* (Madrid, 1788), Part I, p. 100. This is a reference to the fact that Torres, when living in Madrid, completed his first degree and increased his qualifications through the official agency of the University College in Ávila which did not enjoy the prestige of the University of Salamanca.

79 See *Viaje fantástico*, *ed.cit.*, pp. 77 ff.

80 See *Obras*, *ed.cit.*, pp. 254 ff. A paper opposing Torres' findings was announced by the *Gaceta* (21.IX.1751). The prize was awarded to Dr Domingo Talia (*Gaceta*, 5.1.1751).

81 *El ermitaño y Torres*, *ed.cit.*, p. 18.

82 *Vida natural y católica*, 1730, Preface (pp. xx-xxi). The pages are unnumbered.

83 See McClelland, *Benito Jerónimo Feijoo*, pp. 68 ff; p. 118.

84 See *Vida natural y católica*, 1730, pp. 35 ff; pp. 74, 107.

CHAPTER 4

The Psychological Significance of Pulpit Oratory

'. . . how many more people there are that see and hear than think and judge'

(Colley Cibber, *Life*).[1]

Related only incidentally to major surveys of the new meaning of reason in Philosophy and Science, but of prime emotional concern to the *vulgo*, was an intellectual re-appraisal of the quality and manner of religious preaching. Ridiculous as Baroque whimsy in matter, and theatricality in mannerism, might begin to seem to a rationalistic intelligentsia, it had expressed a state of mind as justifiable in its historical context as was that of plain-speaking representatives of Enlightenment. The exteriorization of a complexity of mental bewilderment raised by the Renaissance question 'What is reality?' took various outer forms. One was a display of imaginative suggestion, an elaborate conversion, into ever more elaborate figures of speech, of the epoch's sense of elaborate confusion: a form beloved of literary artists who were not obliged to find definite answers. Another, eventually, was an activity of practical detection, inductive reasoning, and scientific plainspokenness, as exemplified in Bacon, Newton, Descartes, Locke, Leibnitz, Mayáns y Siscar, Feijoo, and their like. Paradoxical ideas, or thought-provoking ornateness, made fashionable by Cervantine or Quevedan genius respectively, were debased by ungifted imitators. Fashionably plain, blunt speech, when used by ungifted men could become, from the literary standpoint, banal and lifeless. But that does not mean that inferior forms of expression, whether inartistically exuberant or banal, were necessarily unrelated to realistic thinking. The conception of a realistic complexity of life is too large for the ungifted many, who vaguely perceive it, to express with either ideological or stylistic coherence. At the same time their struggling understanding, however tediously exteriorized in the confusion of polemic, is an aspect of mental frustration typical of a period made by philosophical and political experience unbearably self-conscious. A crisis occurs when established modes of expression are extended for the purpose of conveying an epoch's revolutionary experience and seem to challenge congruity. The late Victorian conception of reality, for instance, was fundamentally similar to that of the early twentieth century and as well represented. But it was expressed more indirectly and ornately, in keeping with Victorian experience of social inhibitions, and was challenged by a new distrust of outward and dressy appearances. Just such a crisis occurs in Spain's Age of Reason when an established, a Renaissance puzzlement explained itself for too long in a nostalgic language which conflicted with the new terminology used to assess an apparently newer social and intellectual disillusionment. That crisis represents an intelligent phase in the country's awareness of the many-sided possibilities of perceiving and expressing complex reality.

The ornate rhetoric of Baroque preachers in all countries, though apparently at its most extreme in Spain, was outwardly one of the debased expressions of ideological disturbance. But in spirit it was a measure of emotional security, and, like folk-superstition, a defence against the bleakness of unknown or unfamiliar possibilities, the pervasive bleakness of material doubt. When nervous consciousness of the complexity of life was sublimated into an obsession with the details of illustrative imagery for their own sake; where, inside familiarly rounded lines of demarcation, a nervous consciousness intensified evasion of realistic thought, it could preserve an illusion of fulness, completeness, and emotional protection. Pulpit oratory was similar to the oratory of drama. Rhetorical questions, repetition, animating surprises provided by original or unexpected images, ideological paradoxes wrapped comfortingly in multileaved picturesqueness . . ., might appeal only indirectly, if at all, to any intellectual judgment, but could affect the senses directly and immediately. The openminded and free-moving search for the complex meaning of the mysteries of reality in *Don Quijote* or *Hamlet* was ironically oversimplified by lesser writers, who relied on generalization of thought and a sophistication of format to suggest mystery and complexity: a method which elaborately confined ideas within ornately sensual boundaries and was so much easier to understand than disturbing openmindedness. Even Quevedo, who disturbed mental assumptions by shattering his reader's confidence in sense impressions, did so within the limits of highly organized configurations. Cervantes and Shakespeare at their most inspired might depart from pre-conceived guide-lines. *Don Quijote* is inwardly so close to the real, the process-life of mind that it has no true beginning or middle, and an end only engineered for the author's circumstantial convenience. Their successors' questioning and wondering tended to assume a more regular outward shape, and expressed complexity by obsession with illustrative detail or, as in Quevedan satire, engineered an original effect of crisis by a distortion of imagery in which any one sense impression might be expressed in terms of any other. Such highly organized patterns of rhetoric, outwardly logical in shape for all the complexity of their illustrative detail, and therefore giving every outward appearance of soundness and strength, might, when plagiarized by inferior writers, represent a philosophical wisdom which they did not inwardly possess, and, in second-class preaching, might cause preachers and hearers to confuse established method with established truth.[2]

Consequently to break with the outer sophistication of oratory could suggest to either the unintellectual or intellectual *vulgo* a break with theological patterns and principles. To reduce illustrative colour and emotive sound could suggest destruction of the protective armour of Faith. Preachers, like actors, must have been obliged to make their impact on largely illiterate or uncritical hearers by a variety of speech-rhythms and gradations of vocal pitch. The tendency, even nowadays, to achieve sensational effects on the stage by shouting, or by other forms of exaggerated sound from actors or stage machines, is commoner than by a quiet urgency of reasoning. External sound and show are easier to stage-manage than conversational intensity, and more acceptable to the average audience. In certain regions, sects, or groups it still is not unknown for a preacher to drive his hearers into a frenzy of devotion by dint of dramatic speech-techniques. Certain sections of the twentieth-century public are no less dependent than the general public of the eighteenth century on declamatory stimuli. An audience or congregation then, as now, did not and does not need to understand every word of an emotional argument in order to be moved, even converted. A rising pitch of voice, contrived or spontaneous, of itself can induce response. Declamatory repetition and other sophisticated forms of oratory, whether fashionably structured in fashionably picturesque jargon, or fashionably shouted in fashionable disorder, are instinctively or judiciously calculated to hold attention by sound-vibration. Sound can be effective in itself as it is in the rhythms of poetry. To a certain extent, then, the declamatory preachers of the eighteenth century met a social need and were undoubtedly influenced, albeit unknowingly, by their actor-colleagues whom they tended to despise.

As for Quevedan, Torresian or Goya-like elaboration of imagery by sophisticated distortion, that too represented, especially to an intellectual, the reason of unreason, or unreason of reason, which tormented an epoch still unused to envisaging truth as an unknown quantity and still unpractised in suspending judgment. Distortion was a natural reaction to, or development of, over-rational thinking and instinctive acknowledgment that reason cannot explain everything. As a symbol of puzzlement it could suggest the supernatural or the inexpressible, and only failed to suggest spirituality by its lack of simplicity. It often transposed into suggestiveness those stark realities which the average man dared not visualize as such. In general, ornamentation, by the elaborate distortion of incidental imagery, or by the elaborateness of a whole pattern of expression, had become equated, sometimes rightly, usually wrongly, with richness of thought.

In some degree then, actors, preachers and statesmen, until at least the early twentieth century, commonly relied on elaborate mannerisms of word-resonance and fancy figures of speech to claim attention. Neither seventeenth-century theatre audiences nor ordinary Church congregations could possibly have understood every elaborated detail of every whole purport of a Baroque speech or sermon. But theatrical rant and gesture, with its tradition of Senecan declamation informing two centuries of Golden-Age drama, the very pitch and rhythmical movements of the voice, the visual stimulus of attitudinizing, and the sensual lure of concrete illustration and imagery could, even if only half understood, excite, move, convince. And so they could emotionally educate hearers accustomed to such stimuli and to learning or understanding by means of overall impressions.

Intellectuals themselves were usually disdainful of writing that dispensed with studied eloquence. Enlightened as was the Valencian scholar Gregorio Mayáns, and critical as he was of the 'ingenios delirios' (ravings of ingenuity)[3] of the pulpit, he could object to Feijoo's judgment that true eloquence depends not on art but on naturalness, and to the Benedictine's indifference to Classical rules of rhetoric for modern purposes.[4] It should be observed that eighteenth-century critics of Baroque preaching in Spain were not advocating sermons devoid of rhetorical eloquence. They looked to the Classics, notably Cicero, as instanced by his admirer Francisco Isla, who was to caricature contemporary preaching in *Fray Gerundio* under the pseudonym of Don Francisco Lobón de Salazar.

Among Spanish models recommended by this satirist was Luis de Granada,[5] whose admirably Classical eloquence has a poetic quality akin to that of Fernando de Herrera, dependent, like Herrera's, on the rhetorical question, exclamation, ennumeration, judicious imagery, and illustration, and is perhaps more artistic than analytical. From the standpoint of the twentieth century one might have supposed that the men of reason would have preferred to Granada's formality the stimulating spontaneity of, for example, St Teresa of Ávila, who, as a spiritual model, was certainly well known to the eighteenth century. Calculated eloquence, however, continued to serve as the sacred dress – the speech surplice – in which serious thought, or what passed for serious thought, ought to be presented and recognizable as such.

Renowned by these men of reason as a Granadan model near their own time was the Portuguese Jesuit Father Antonio de Vieira, whom Mayáns recommended to student preachers and whom Isla also praised warmly. A series of his sermons for Lent had been published in its Spanish edition in Madrid in 1711 as *Las cinco piedras de la honda de David*[6] and acted as useful reference for reformers. Yet Vieira, who died in 1697, had inherited to a certain exent the Baroque habit of equating depth of thought with complexity of reasoning and detail, and of expressing inner complexity by outer complexity of judicious word-play or concrete imagery. What rightly impressed reformers in Vieira was his attempt at analysis – his treatment, for instance, of the meaning of self-knowledge and man's methods of obtaining it. In general Vieira neither indulged in festive oratory for its own theatrical sound nor allowed himself to be guided by current tendencies to describe rather than to examine:

> Si le digo, que se conozca por la parte inferior y terrena, temo que un concepto tan bajo de sí produzca acciones viles como en Adán; si por la parte superior, y tan alta, temo que la misma alteza de su conocimiento degenere en hinchazón, y soberbia, como en Lúcifer . . . Hombre si te ignoras, si no te conoces a ti mismo, sal afuera.[7]

> (If I tell him to know himself on the lower terrestrial level then I fear that so low a conception of himself will produce the vile actions produced in Adam; if on the superior and so high a level, I fear that the very height of his knowledge will degenerate into swollen headedness and pride as in the case of Lucifer. Oh man, if you are not aware of what you are, if you do not know yourself, then get away [from yourself].)

But traditional over-emphasis on the form – as distinct from the aim – of Scholastic reasoning caused him to search too extensively for unusual argumentation and so, often enough, to appear more ingenious than profound, more rhetorically illustrative than spiritually suggestive:

> De manera que, caminando del bien a la pérdida y de la pérdida al dolor, el bien, la pérdida, y el dolor son menores; pero volviendo de la pérdida al bien, y del bien perdido al dolor, el dolor, la pérdida y el bien son mayores; y todo esto siendo el bien el mismo, y no diverso . . .[8]

> (So that moving from what is gained to what is lost and from what is lost to what is suffered, the gains, the loss and the suffering are lesser evils; but returning from what is lost to what is gained, and from lost gains to suffering, the suffering, the loss and the gains are of major importance; and all this according to the fact that the gains are of one kind and not diverse.)

As Maimó y Ribes reported from 'El Barbadiño''s respectful survey of Vieira, he unfortunately set the kind of example of subtlety which destroys eloquence.[9]

One of the wisest and most succint pieces of advice to preachers was given by a contemporary of Vieira, the Archbishop of Toledo, Francisco Valero y Lossa who died in 1720. Significantly, he took for granted the influence on public speakers of the theatres, and was at pains to distinguish between subjective ideals of showmanship and objective ideals of instruction:

> El que lee de oposición en el Teatro, procura decir cuanto puede, y cuanto sabe, entiéndalo quien lo entendiere; pero cuando se lee en el Aula a los discípulos, procura el maestro acomodarse a la capacidad de los oyentes: porque los fines son distintos, el primero de lucir, y el segundo de aprovechar; sea pues el púlpito Cátedra de Aula, y no de Teatro.[10]
>
> (He who speaks his lines competitively in the theatre, endeavours to say as much as he can and knows, whether everybody understands it or not; but when the master speaks his lines to his pupils in the lecture-theatre he tries to accommodate himself to the capacity of the hearers: because the aims are different, the first that of showmanship, and the second that of usefulness; therefore may the pulpit be the Lecture-room Chair and not that of the Theatre.)

Patterns of mental expression do not automatically change to suit the political turn of a century. Few preachers there would be in Spain or abroad who would deny themselves an occasional parade of rhetorical culture if only for the emphasis of conclusion, like this finale of Vieira:

> Aplicad, y meted estas cinco Piedras en aquellas cinco Fuentes de Misericordia. Teñidlas, y bañadlas muchas veces en el torrente de aquella preciosísima y potentísima Sangre, porque bañadas en aquel torrente, y en aquel torrente purificadas, suplirán abundantisímamente mis defectos y serán *Limpidísimas lápidas de torrente.*[11]
>
> (Apply (yourselves) and put these five Stones in those five Fountains of Mercy. Dye them and wash them over and over again in the bloodstream of that most precious and most powerful Blood, because, washed in that stream and purified in that stream they will replace most abundantly my defects and will be *utterly clean stones of the bloodstream.*)

In France even the revered Bishop of Clermont, J.B. Massillon, whose sermons, acceptable as models of discreet elegance, avoided figurative and rhetorical exaggerations, obeyed at times a folk-instinct to indulge in vocal histrionics: the trick, for example, of gradually engineering a pitch of excitement by repeating key-words emphatically in different, or significantly identical tones, until they induce a climax-shout:

> Heureux le prince, vous dirait-il, qui n'a jamais combattu que pour vaincre, qui n'a vu tant de puissances armées contre lui que pour leur donner une paix plus glorieuse, et qui a toujours été plus grand ou que le péril ou que la victoire! . . . Heureux le prince . . . Heureux . . . non celui . . . non celui . . . non celui . . . non celui . . . non celui . . . etc.[12]

Perhaps the very tradition of arranging sermons in clearly defined Parts or *Puntos*, like layers of rising suggestiveness, with predisposing Introduction and catharsis-style finale or Conclusion, encouraged preachers to develop theatrical techniques. This traditional arrangement is warmly approved by a sponsor of Alonso de la Guardia's emotive three-Part *Oración funebre panegírica* for D. Juan de Cereceda, preached in 1743.[13]

If Isla's dutiful praise of a French Jesuit reformer, his near contemporary, Father Blaise Gisbert, rings a little less enthusiastically than his praise of Vieira, that would not be because of Gisbert's conception of preaching as a highly refined, practical technique of verbal sublimity achieved by efficient organization, in logical sections, of ideas and judicious dialectic. Gisbert's *L'Éloquence Chrétienne dans l'idée et dans la Pratique*, first published in 1714, republished in 1728,[14] and translated into English and Italian, had become a textbook for students of oratorical reform. Even a Protestant theologian, such as Jacques Lenfant could express approval of Gisbert's conception of oratorical eloquence, only railing at his tendency to overstate his case:[15] a point duly mentioned in *Fray Gerundio*.[16] But Isla's general respect for Gisbert must have been restrained somewhat by the Frenchman's denunciation of satire in clerical mouths:

> Éloignez de vos discours avec un soin scrupuleux tout ce qui a quelque air, quelque apparence de satire . . . On ne monte pas en Chaire pour contenter ce penchant naturel à l'homme, de censurer . . . On ne persuade pas un homme en se moquant de lui et de ses défauts, ou en lui disant des injures. Il ne faut qu'un trait malin et satirique qui vous échappe, pour révolter tout le public contre vous . . .[17]

Though Gisbert here was referring directly to pulpit oratory, the tone of the context suggests that he would have been no friend to clerical satire in print. On the other hand Isla, together with many pulpit reformers, Spanish or otherwise, might well have been remembering Gisbert's words in another context when they criticized preachers and public for regarding the pulpit as a stage:

> Tout ce qui tient de la représentation est de leur goût. Les gens d'esprit vont a leurs sermons comme à un spectacle.[18]

In this respect it is notable that an English translator of Gisbert, Samuel D'Oyley, a 'late Fellow of Trinity College, Cambridge', had observed a curious difference between the reactions of English congregations and of French congregations, especially when the latter were addressed by Jesuit preachers, in that Latin Churchgoers expected fire in the speakers' voice and demeanour and felt free to react with 'applause' or 'murmurs':[19]

> Without a great deal of fire in the speaker the finest Discourses would be insupportable to them; and for this reason, I believe, the Jesuits who study human nature very diligently, discover more earnestness of expression, and more variety of action than any other preachers, that I have observed among the Catholics. Indeed they carry this point to a great extreme, and how

agreeable soever their extravagancy may be to the multitudes that follow them, it is very shocking and ridiculous to others.[20]

Why, then, should a reformer in Latin countries fail to understand that sermons were necessarily regarded as theatrical show? Indeed, to judge by Jonathan Swift, even D'Oyley's 'passive' and 'silent' congregations in England expected their preachers to declaim with some dramatic effect.[21] The folk-need of drama in one form or another would become more logically purified only when recreational facilities became more sophisticated.

So enlightened preaching was not designed to break with traditional forms, but to refine and moderate them and to improve intellectual content. One of the many criticisms of Feijoo's impatience with preachers liable to 'perder el tiempo en florecillas inútiles' (lose time over useless little flowerinesses)[22] was that his stylistic unconventionality detracted from the dignity of his professional status.[23] Surviving examples of Feijoo's sermons are characteristically simple in language by contrast with other sermons preserved in the library of his monastery, and so convinced did Feijoo become of the need for naturalness in the pulpit that he accused himself of having sometimes sinned by ignoring such a need.[24] Isla, on the contrary, frequently resorted to the heavy artifice of virtuosity, particularly in pamphlets, even in private letters, to such an extent that, unlike Feijoo, he was not always aware that he had sinned.[25] His own sermons were expressed in general with progressive straightforwardness between, for instance, 1728 and 1736, perhaps in response to the writings of Feijoo. They show a healthy tendency to an individual analysis of moral questions. Yet they are not devoid of stylistic artificiality. Emphasis, especially in the preacher's summaries, still depends on the declamatory technique of repetition and enumeration:

¡Gracias a Dios que una vez vemos al mundo tratado como merece! ¡Gracias a Dios que se hace de él la estimación que le corresponde, y el aprecio que se le debe! Dignidades, empleos, nobleza, honores, opulencias, diversiones, profanidades, faustos, alegraos, regocijaos, digo, que ya tenéis restituido el crédito que se os había quitado: ya se os representa con el semblante que tenéis, y no con el que se os fingía: ya os dejáis ver como sois, y no como os figuraban . . . ¿Y esto por qué? Porque se hacía burla de ellos, eran la mofa, el escarnio, la irrisión de todos. De todos, digo . . .[26]

(Thanks be to God that for once we see the world treated as it deserves. Thanks be to God that there is made of it the estimation which corresponds to it and the valuation due to it! Dignities, employments, nobility, honours, opulences, entertainments, profanities, pomps, gaities, merriments, I say, now there has been restored to you the credit that had been taken away from you: now you let yourselves be seen in your own image, and not in the imaginary one . . . And why this? Because one made fun of them, they were objects of mockery, the derision, the laughing stock of all. Of all, I say . . .)

The inspiration of Quevedan paradox hovers here and there:

> Dice San Gabriel, que la virtud del Altísimo hará sombra a María . . . Quiere decir, exponen los dos Thomases, el de Aquino y el de Villanueva, ambos lucidísimos Soles de la Iglesia, que pueden hablar bien en materia de luces y de sombras: no que el Sol de Justicia Christo había de asombrar u oscurecer a María, sino que María había de ser sombra del mismo sol . . . Ahora, . . . hay sombras de sombras . . .[27]

> (St Gabriel says that the virtue of the Most Highest puts Mary in the shade . . . That is to say that the two Thomases, he of Aquinas and he of Villanueva, both most brilliant Suns of the Church, show that they can speak well on matters of light and shade: not that Christ, the Sun of Justice was to overshadow or obscure Mary, but that Mary was to be the shadow of the very sun . . . Now . . . there are shadows of shadows.)

A critic's acceptability has much to do with his manner of criticizing, whether that be calculated or instinctive. Some of the reformers were reasonably objective and criticized with understanding of the difficulties involved in changing habits which the average man regarded as sacred. These reformers generally wrote unobtrusively, through academic media, without aggressiveness and without engaging in personalities: a method which certainly encourages serious discussion and has a gradual, pervading influence. It does not, however, produce rapid results. The unimpassioned Mayáns and the Jesuit, M.S. Burriel, quietly corresponding on the subject, assumed that reform was difficult. And Burriel, while attempting in general to bring his own sermons up to the standards of judicious eloquence recommended by Mayáns, protested that a style emptied of all imagery was regarded as work on which no effort had been expended. He could quote cases of the hostile reception given to plain sermons by both clergy and 'vulgacho' (lowest of the low). But his strongest evidence of the difficulty of reform lay in what he called his lack of courage: the fact that, when asked to preach for special occasions, he was unable to follow his better judgment and go against the tide of public opinion.[28]

The legal mind of Mayáns, Professor of Law at the University of Valencia, was astute enough to be able to detach Vieira's analytical content from his 'alluring' style, and he noticed that Vieira's admirers tended to be unduly influenced by the latter 'wherein they were unable to sustain themselves'.[29] To his credit, too, he admired the writings of St Teresa of Ávila, though more, one suspects, for their content than for their unpretentious expression. His own style in his *Orador cristiano*, a 'Dialogue' in Classical vein designed here for uncompromising instruction rather than for pedagogic discussion, is what he understands to be 'familiar, pero sin bajeza' (familiar but without vulgarity).[30] Perhaps we should now call it academic. Certainly his *Orador* is concerned as much with matter as with method. The preacher, says Mayáns, must be a 'very rational logician'.[31] He should interpret biblical texts without constant recourse to metaphor, rhetoric, ridiculous gesture or theatrical tone, and without exaggerated condescension to local interests. At the same time Mayáns' insistence on the essential requirement of Dialectic training, the importance of studied eloquence as exemplified in Classical models, and the need of 'Art' to replace the holy inspiration of the Apostles, shows that his use of words like 'naturalness' or 'popular' cannot be interpreted in a twentieth-century sense, or even in the sense in which they may be applied to Feijoo, of whose conversational style and occasional colloquialisms he so earnestly disapproved.[32] Which means that Mayáns, like Vieira, Isla and the *vulgo*, assumed the need of vitalizing rhetoric, while differing from the *vulgo* in his recognition of rhetorical abuse and degeneracy.

Another reformer of Church oratory, the Capuchin Padre Fray Matías Marquina, elaborated even more systematically on the need to return to the seeming simplicity of Classical refinement. Like Mayáns' *Orador*, Marquina's *Escuela general histórica* of 1751,[33] a collection of sermons ambitiously dedicated to the Queen, sets a good example regarding the subject-matter of sermons: his analysis of St Anthony's moral principles, for instance; his call for spiritual development; his insistence that the major function of a sermon is to direct men to virtue and dissuade them from vice. These sermons, however, were composed for the purpose of giving instructions not merely, not even chiefly, on content but on style. To Marquina as to Mayáns, Isla, Vieira, and their reformist compeers, a sermon was regarded as a piece of literary creativeness, as if in competition with literary works of art. Presumably the literary wealth of the Golden Age had influenced the pretensions of all public writers and speakers and, since all public speakers cannot attain to those high standards, the very wealth of Golden-Age achievement became indirectly responsible for the wealth of third-rate imitativeness. So the Marquinas speak more of the preacher's method and less of his realistic observation of details of human behaviour. Here is the traditional play of paradox and verbal parallelism, sound in moral intent, but too self-consciously artistic:

> . . . si resucitas mal, de nada sirve el haber nacido bien, porque la mayor gloria de la vida consiste en resucitar a la vida de la gloria.[34]
>
> (. . . if you resuscitate badly, there is no point in your having been born well, because the greatest glory of life consists in a resuscitation to the life of glory.)

Here is the typically reformist call for the improvement of rhetoric as a powerful means to a vital end. Rhetoric rightly organized – the emphasis is on organized technique –, sensitively proportioned and discreetly applied, signified power and force, the essential means of firing men's will:

> Esto debieran considerar los que se desvelan por regir hombres sin hacer aprecio de regir sus voces . . . Esta es la causa de que habiendo tantos Predicadores en el mundo, haya tan corto fruto evangélico, pues debiendo sazonar la tierra con su doctrina, erudición, y elocuencia, la desazonan, y esterilizan con la falta de la sal, que es la Retórica.[35]

> (This ought to be taken into account by those who are so overconcerned about directing [the behaviour of] men and yet who do not [rightly] direct their own voices . . . This is the reason why, though there are so many preachers in the world, there is so little evangelical fruit, for while they ought to be seasoning the earth with their doctrine, erudition and eloquence, they take away any taste and sterilize the effect with their lack of salt, which is Rhetoric.)

No wonder that the conversational abruptness of Feijoo should have seemed to Mayáns, among others, so disturbingly ungainly.

Incidentally it is interesting to observe Marquina's formulation by stylistic means of a natural law of moral criticism which Golden-Age playwrights, for instance, had used with conscious or unconscious insinuation. A safe way of rebuking and criticizing social superiors – princes and high ranking officers of State – was to praise them for virtues which notoriously they did not possess. As Marquina delicately specifies:

> Varios modos de iluminar tiene la Retórica para desengaño de los Príncipes, o bien declarando las virtudes que le deben adornar, para que las solicite, o bien panegirizando las que otros de su misma esfera tuvieron, para que las emule.[36]

> (Rhetoric has various illuminating ways of undeceiving Princes, either by emphasizing virtues, which ought to adorn a Prince in order that he should desire them, or praising those which others of his own sphere had possessed so that he might emulate them.)

For psychological reasons it would take time before rhetoric could be partially laid aside, much less generally discredited. Change in political systems would cast disillusionment over inherited magnificence of speech. The temporary fall from international power of the Society of Jesus would contribute to the fall from favour of Scholastic organization of argument. Empiric technique in laboratories would encourage straightforward reporting. In the meantime, however, that is, during at least the first half of the eighteenth century, rhetoric in public speaking was an understandably natural requisite.

The difficulty of reforming pulpit histrionics, then, was an accepted fact of intellectual life. Even so, it is possible to offend without outrage. The blunt tones of Feijoo or the academic arguments of Mayáns rang with a disinterested sincerity which produced more annoyance than scandal. But when in *Fray Gerundio* Father Isla, aggressively conscious of current criticism against his Order by other Orders and by the secular clergy, resorted to interested specifications of pulpit vice, he inevitably provoked outrage. Allegations about Jesuit assumptions of intellectual superiority belong to the religious and political history of the eighteenth century and need not be discussed here. It should be remembered in general, however, that the Society, in the eyes of its critics, was secretively dedicated, as a Spanish document of the period puts it, to 'maintaining the glory of its own initiates'.[37] Incidentally, therefore, the unpopularity of the Order, for whatever general reason, tended to affect the reception of any criticism on any subject by any of its members. Nor did Isla, as an individual, commend himself easily.

He was not among the most modest of his Order. His letters and polemical articles before the Expulsion smack of personal conceit and self-righteousness.[38] For other reasons, too, he was an intellectual liable to be misunderstood. He was only moderately modern in academic outlook; dubious of the claims of experimental scientists;[39] prejudiced against those who questioned the permanent value of traditional Scholastic techniques – still followed, he noted approvingly, by certain members even of the Anglican persuasion such as George Bull.[40] Also he was inclined to select his evidence to suit his convenience and to indulge in criticism too sweeping for accuracy. One of his critics, the rational José Maimó y Ribes, who applauded *Fray Gerundio* but objected to certain details, especially Isla's incidental attack against the Portuguese theologian El Barbadiño, instanced as incongruous, in an enlightened scholar, the assertion made by Isla that Antonio Gómez Pereyra had anticipated Newtonian science. Here Maimó y Ribes with Feijoon directness underlines Isla's limitations:

> En esto me parece padeció V.md un grave descuido, pues quien profiera semejante proposición da a entender que tiene poca o ninguna noticia de la Filosofía, y Física moderna, y V.md la tiene muy extensa. Los filósofos modernos de juicio siguen un sistema que es no tener sistema: nada creen, sino aquello que se les prueba con evidencia; y esta prueba se alcanza con las experiencias repetidas, que hacen ver y descubren la esencia de la naturaleza, sus propiedades, efectos, y demás.[41]

> (In this respect it seems to me, Sir, that you make a serious slip, for the man who offers such a proposition indicates that he has little or no knowledge of modern Philosophy and Physics, yet you, Sir, have a very extensive knowledge of these subjects. The modern Philosophers of good judgment follow the system of having no system: they believe nothing but what can be proved to them with evidence; and this proof is obtained by means of repeated experiments which demonstrate and disclose the essence of nature, its properties, effects, and the rest of its qualities.)

'. . . the system of following no system . . .' that is what Isla and most enlightened Jesuits, despite themselves, found most difficult to understand.

Nevertheless Isla was openminded enough to admire Feijoo in general and publicly to support him.[42] He was the middle-way scholar, mentally healthy because mentally curious, both in his doubts about the progress of reform and in his convictions that reform was necessary, but the kind of thinker most open to misinterpretation by reformists and anti-reformists. When to this fact is added Isla's conspicuousness as a Jesuit in times of Jesuit unpopularity, and the acidity of his satire, it is no wonder that, even with the best intentions, he could create alarm and so polemical upheaval. However altruistic in purpose reformers may be, they are not invariably free in practice from personal pride and subjective ambition. Isla is so thoroughly carried away by the exhilarating force of his own creativeness that he allows the creatures of his caricature to take full charge: an artistic triumph indeed, but one which, in his circumstances, transgressed against propriety and religious charity. His talent, in Quevedan and Torresian style, for providing inanimate objects with animating qualities, such as 'catarrhal' bells,[43] or the originality of his irony – his jeers, for example, at foreign arrogance: '. . . hablo en país libre; que en Inglaterra yo me guardaría de hablar de esta manera' (. . . I speak in a free country; for in England I should take good care not to speak in this way)[44] – make him immortally readable. But his bouts of irrelevant coarseness, of excrement and backside association, also in Quevedan and Torresian tradition, and figuring not only in *Fray Gerundio* but in articles and personal letters,[45] were as little becoming to his Jesuit cloth as was the ingenious coarseness of his brother satirist, Jonathan Swift, to the cloth of an Anglican Dean.

Much of Isla's artistic licence must have been used unconsciously, or, in his schizophrenic composition of priest-artist, as the temporarily alienated expression of artist-personality. In his other character as priest and religious he had solemnly preached a sermon, later printed under the title of 'Sermón de la corrección fraterna', which compassionately outlines the accepted procedure for the denunciation of religious faults. That procedure, we notice, does not include exposure to public ridicule:

> . . . mirad, si alguna vez viereis pecar contra vosotros alguno de vuestros hermanos, id, y corregidle a solas por la primera vez: a la segunda, corregidle delante de uno, o de dos testigos; y a la tercera si no se enmendare, delatadle a la Iglesia . . .[46]
>
> (. . . look, if on any occasion you should find one of your brothers sinning against you, go privately to him the first time and correct him; on a second

occasion correct him in front of one or two witnesses; and on a third occasion if he makes no amends report him to the Church.)

Yet he was not a man to underestimate or be unaware of the nature of his own talent, and, when roused by artistic interest, he both instinctively and calculatedly followed his inspiration to the end, extreme as that end might seem. His belief that his *Fray Gerundio* was an eighteenth-century parallel to *Don Quijote*,[47] and himself Cervantes' artistic equal, strengthened his sense of righteousness in permitting himself such freedom of satirical movement. But beneath the surface of *Fray Gerundio* there lurked a destructively jealous motive which, even had Isla possessed the genius of Cervantes, would have rendered his novel inferior to *Don Quijote*. Much as he might fancy himself as an eighteenth-century Cervantes, Isla was not wondering, not satirizing openmindedly. In a prevailing atmosphere of animosity among the Orders he was in part directing his criticism against those Orders which were suspicious of Jesuit domination, those less ready to expand intellectually in a Bourbon climate than were the Jesuits, those orders, therefore, which received wider popular sympathy. Incidentally, therefore, he helped to inflame anti-Jesuit propaganda. Not merely because *Fray Gerundio* satirized bad habits of preaching in general was it banned by the Inquisition two years after its first appearance in 1758, and banned with its sequels in later years. From the religious standpoint the truly censurable feature of the work was its tone of malice which, if in the first volume it had hovered on the level of implication, was raised in the second volume to the level of dominance.

It is true that Spanish critics of the period, in the context of world-power and of a national sensitivity to political defeat, were more liable to constrict their sense of humour than in times of greatness. In Isla's jaundiced view, as expressed in a letter of 1763 to Langlet, who had mooted the possibility of publishing a Spanish equivalent of the Jesuit Journal of Trévoux, Spaniards understood only praise or blame and were therefore incapable of supporting the free-range criticism of such a journal as the *Mémoires* of Trévoux:

> Nuestros autores no entienden *raillerie*, ni mucho menos nuestros *autorcillos*, que en España, como en todas partes, son en mucho mayor número. O se les ha de alabar, o no se les ha de contradecir. No reconocen otro tribunal para juzgarlos, [i.e. enlightened critics] que el de la Fe y el de las buenas costumbres y regalías. Niegan la jurisdicción de la crítica . . .[48]

> (Our authors do not understand *raillery*, and much less do our *minor authors* understand it, and they in Spain as in other parts are in the majority. Either one has to praise them or to fail to contradict them. They do not recognize any other tribunal for the judgment [of enlightened critics] than that of the Faith and of good customs and Royal prerogatives. They deny the jurisdiction of criticism.)

Isla here was wrong only in his chronological generalization, in that he applied to Spaniards of all types and times the sensitivity of eighteenth-century nationalists. He was right in that his own satire could not be viewed objectively. The very fact that he expressed embarrassment at the thought of the low opinion which foreigners, especially Frenchmen, must be receiving of the intellectual standards of Spanish clergy, initially alienated many readers who otherwise could have laughed naturally at Fray Gerundio's absurdities. Similarly, since it is so much easier for a country to laugh at itself when it can take its greatness or importance for granted, as in the times of Cervantes or even Quevedo, Feijoo's admiration of various foreign masters and his criticism of his compatriots appeared, for emotional reasons, ill-timed.

Such sensitive reactions apart, there are two ways in which criticism made or implied in *Fray Gerundio* would be likely to affect Spanish readers: ways corresponding respectively to Isla's dispassionate moods and to his moods of personal malice. In Spain and elsewhere professional, even theological, objections to current standards of preaching were voiced by clerics and laymen and, where quietly and disinterestedly reviewed, had not necessarily offended men of reasonable intelligence and honesty. The movement to promote a plain, strictly accurate speech of reason is observable in Spain not much later than in other countries. Its Spanish representatives – Feijoo, Mayáns,[49] Isla, Climent[50] etc. – complain of features of pulpit oratory, showmanship, and general ignorance similar to those condemned by their clerical colleagues abroad.

Swift, when attacking religious abuses in his *Tale of a Tub*, of 1704, refers to preachers' 'wind' and 'oracular belches', to the man who has forgotten 'the common meaning of words' but remains 'an admirable retainer of the sound', and to the 'whole operation of preaching . . . to this day' as being 'styled by the phrase of *holding forth*'.[51] The *Spectator*, which had talked disparagingly of our 'Upper-Gallery audiences . . . who like nothing but the husk and rind of wit, prefer a quibble, a conceit, an epigram, before solid sense and elegant expression . . .', later turned his attention to pulpit oratory: 'The Dissenters . . . do indeed elevate their voices, but it is with sudden jumps from the lower to the higher part of them, and that with so little sense or skill that their elevation and cadence is bawling and muttering . . .'.[52] John Wesley, prefacing a collection of sermons published in 1788 decried 'ornamental' style and 'French oratory',[53] since his reformist aim had been to shatter preconceptions and tear away the comfortable jargon in which religion had wrapped itself. All the same, he had himself indulged in some hell-fire rhetoric in his call for mercy at the end of his Sermon in Three Parts: the 'Cause and Cure of Earthquakes', of 1750, where a lengthy interplay of interrogatives and imperatives promotes a stage-tone:

> Had we no reason to expect any such calamity? No previous notice? No trembling of the earth ..? No shock . . .? Had we never heard . . .? . . . I warn thee once more, as a watchman over the House of Israel, to flee from the wrath to come . . . Art thou ready to die? . . . Believe . . . Believe . . . Come to the friend of sinners . . . Ask and thou shalt receive . . . Call upon Him . . . Wrestle for the blessing . . . Cry mightily . . . etc., etc.[54]

And to produce the mass hysteria in his own congregations, which later worried him, he must, one supposes, have had recourse to many theatrical tricks of intonation.[55] Likewise his brother Charles knew, from theatrical tradition or folk-instinct, how acceptably to present plain ideas in a fury-dress of melodrama. There is more than one form of ornamentation. An elaborate emotionalism of sound, in direct speech, achieved by calculated patterns of exclamatory, imperative and interrogative clauses – however straightforward each individual clause is in itself – can produce as complex an effect as Baroque imagery. On the part of congregations the tendency in each case was to respond to the general effect rather than to absorb every idea or the meaning of every word. Dr Johnson was reported to Boswell as having observed 'that the established clergy in general did not preach plain enough; and that polished periods and glittering sentences flew over the heads of the common people, without any impression on their hearts'.[56] While Voltaire, critical though he was of exaggerated oratory in the pulpit, more perceptively recognized, like many critics of Isla, that public reaction to interference with traditional systems of communication is one, inevitably, of violent shock.[57] So that change, we must understand, might best come about through persuasive gradualness.

Yet Spanish ridicule of pulpit inadequacies and pretensions in the eighteenth century was not inspired entirely by foreign satirists, though it was encouraged by their publications. Just as the ludicrous inadequacy of medical practitioners had stimulated Spaniards' sense of irony and provided Golden-Age dramatists with themes of foolery, so, prior to the eighteenth century, the ludicrous exaggerations of oratory in Church or elsewhere had not passed unnoticed, even if irony entered Church premises with caution. In *El mejor alcalde, el rey*, Lope, through the muddled reporting of clerical teaching by the *gracioso* (fool), seems slyly to be laughing at the kind of mispronunciations and biblical confusions made by humble *curas* (parish priests)[58] as instanced later by Feijoo and Isla. More overtly Quevedo, or his contemporary Pedro Espinosa, for the authorship seems uncertain, had had the irreverence to create a coarsely spoken preacher, the 'Doctor Sumo Campo Loco' (Doctor Super Wild

Rurality) to satirize, with Quevedan relish and abandon, the vices of his congregation in *El perro y la calentura*. This conceptist paradox of vice rollocking out of holy mouths to land laughingly in sacred precincts was as indecorous in its way as was Isla's choice of his medium Fray Gerundio, who profited from the tradition of all-embracing burlesque. A second edition of *El perro y la calentura* had been published in 1736 and is praised by one of Isla's supporters as good Quevedan satire.[59] Yet such examples, far from suggesting active disillusionment with old habits, rather attested to good humoured acceptance of comfortably familiar ironies. There is an emotional difference between family or friendly teasing and coldly intellectual criticism from without. Isla stood on a satirical borderline, which, in the opinion of the average observer, he frequently overstepped.

To a certain extent, then, he was adopting in content an acceptably national practice. To a certain extent he was violating, by his tone, acceptable social sentiments and conventions. His objection to widely known instances of clerical ignorance was, from a modern standpoint, reasonably emphasized. In parallel examples, mimicking much unhistorical supposition by rustic clergy, he made Judas a gardener in Pilate's house. He included absurd anachronisms – the Virgin Mary learning the 'Ave María' at her mother's knee. He caricatured parsonic jargon or undue rusticity of expression; stagey rhetoric inviting gesturing and general affectation in the Senecan tradition; misleading illustrations from fables, popular proverbs, and comic stories; the forced relating of religious truths to objects of local pride; the twisting, therefore, of religion to suit local convenience. And he ridiculed the condoning of certain religious festivities by the clergy which, however innocently, bordered on heresy when degenerating by popular custom into drunken merriment:[60] a custom that also had shocked Feijoo.[61] Reasonable seems Isla's argument, nowadays, that there is a difference between the sound use of Dialectics for the purpose of organizing theological truths, and ignorant imitation of dialectical technique for the purposes of pomp and show. If people regard sermons as forms of entertainment, aptly wondered the Jesuit, how can they learn from them seriously to repent their ways?[62]

One of the most obvious examples of degenerate orators at the time was the Franciscan P. Fray Francisco de Soto y Marne, who in 1738 had published a collection of his sermons with the Prefatory assurance to his readers that 'todos los sermones que te ofrezco en este tomo, los he declamado en el pulpito' (all the sermons that I offer to you in this book, I have declaimed in the pulpit).[63] The very title of this masterpiece of degeneracy: *Florilogio sacro. Que en el celestial ameno frondoso Parnaso de la Iglesia, riega (místicas flores) la aganipe sagrada fuente de Gracia y Gloria Christo con cuya afluencia divina incrementada la excelsa palma mariana . . . (Sacred Florilogue. Where, in the celestial, charming and leafy Parnassus of the Church, that sacred Aganipe Fountain of Christ's Grace and Glory, with whose increasingly divine affluence of sublime Marian palm . . . sprinkles* [*mystical flowers*]) sets the tone for the entire collection. Every title of every individual sermon competes in cheap paradox with theatrical fashions. There is even a 'Descender para subir' (*Descend to climb*) to compete for publicity with the title of Moreto's religious play *Caer para levantar* (*Fall to rise up*), and a 'Muere de lo que vive, quien vive de lo que muere' (*He dies through living who lives through dying*) aping, perhaps, the *Morir a un tiempo y vivir* (*At once to die and live*) of Juan Cabeza's drama.

The very Dedication to that 'Vice-God of the World', Saint Joseph, in some fourteen fulsome pages of linguistic juggling, prepares a way for the fancifully dubious theology of, for instance, the Franciscan's eulogistic sermon on San Andrés, in which he explains that Christ's outward fear of the Cross was a form of divinely inner humility designed to permit the external fortitude of St Andrew in similar circumstances to appear in greater glory. Contrived interpretations of Scripture for the greater glory of local or popular saints or for great religious festivities, however well substantiated by sanctified authority – like the authority of St Bernardino of Sienna for the interpretation of the 'other angel' of the Apocalypse as St Francis – were dangerously easy to emulate. Parades of classical culture incorporating such phrases as 'Júpiter divino',[64] elaborate comparisons of Venus arising from the sea with the Immaculate Conception of the Virgin, and the splendour of sainthood with the all too obvious brilliance of sun, moon and stars. And the incessant use of senseless superlatives and sensual ornamentation conformed to common standards of preaching technique. Soto y Marne represents the mass of sermon-makers. His innumerable colleagues in that profession, academic and non-academic, brandish their own varieties of the same characteristics. There are the concretely expanded interpretations of biblical stories for modern local purposes, as in, for example, Miguel Abio y Costa's sermon on the new church of Nuestra Señora del Pilar in Zaragoza, whose construction had been delayed but whose completion in the name of Pilar meant that now 'podemos conseguir de su piedad los favores que a título de Madre da a entender no los quiere Cristo Señor Nuestro conceder' (we can obtain through her piety the favours which under the title of Mother she can concede when our Lord Christ may not wish to concede them).[65] For Christ, says Abio y Costa, though initially unwilling to turn water into wine at the wedding feast He attended, had eventually yielded to His

Mother's pleading, choosing, not crystal containers for that miracle, but stone jars – or *pilares*: a word supplying Abio y Costa with a theatrically dubious means of sanctifying local pride. There are the elaborate illustrations of a Fr. Juan Rumualdo Agramonte drawn from a mixture of Classical and biblical sources to impress his hearers with the preacher's scholarship.[66] There is the descriptive oratory of a Fr. Alejandro de la Concepción, one of whose sponsors, Fr. Alonso Cano, thinks that in disapproving the 'tropiezos de autoridades, analogías, delirios mitológicos . . .' (stumbling of authorities, analogies, mythological ravings . . .) he is strengthening the opposite values of Fr. Alejandro's pedantic 'Oración Hortensiana',[67] in which the preacher so often seems incapable of using a noun without the accompaniment of high-sounding adjectives, and expends his energy in vocative address, as in the first two pages of his recorded sermon. There is the vacuous raving of a Fr. Alonso de Huecas, one of the most artificial of sermon-mongers, with its flowery preliminaries to discussion of the virtues of St Francis:

> Alarga la mano a la *rosa*, y la *violeta* igualmente le atrae, para elegir. Aparta de la *rosa* la mano, extiende a la *violeta* su impulso, y le cecea el *narciso* . . .[68]
>
> (He stretches forth his hand towards the *rose* and the *violet* attracts his choice equally. He draws his hand away from the *rose*, turns his attention to the *violet*, and the *narcissus* lisps across to him.)

the insensate word-play:

> Si al que dan en qué escoger, le dan en qué entender, nunca tuvo más en qué entender mi ignorancia; porque nunca tuvo más en qué escoger mi insuficiencia.[69]
>
> (If to the one to whom they give a means of choosing, they also give a means of understanding, then, never did my ignorance have more to understand, because never did my insufficiency have more from which to choose.)

and self-consciously Scholastic exhibitionism. There is the vocative thrust of sentimentality in funeral orations such as the *Llanto universal de la Montaña, sentidos ayes, y tristes balidos de sus ovejas. Por la muerte de su venerable, y ejemplar Pastor, el . . . Obispo de Jaca* (*Universal Tears of the Mountain, heartful cries of woe, and sad bleating from the sheep. This for the death of their venerable and exemplary Shepherd, the . . . Bishop of Jaca.*);[70] of Orencio de Bergua whose flair for elucidating pathos histrionically was fashionably overdeveloped. Or, set firmly in common usage, there are the fulsome panegyrics declaimed by such as Alonso de la Guardia, whose funeral oration of D. Juan de Cereceda, Caballero del Orden de Calatrava, obtains its sob-effects by a throbbing beat of disturbing words: 'murió . . . murió . . . murió . . .' (he died . . . he died . . . he died . . .),[71] and from obsessively illustrative details of the dead man's lowliness and charity. These are the stylistic qualities which rightly provoked Feijoo's outrage and would seem to justify all the wealth of satirical contempt of which Isla was capable. One might add that the preachers' common resort to miracle-narrative also acted as a form of stylistic colour, and, in its emphasis on externals rather than on inner mysteries, would have encouraged those Jansenist tendencies in reformers discussed by Mestre Sanchis.[72]

Nevertheless, just as public circulation of the word 'scepticism' had distressed many who saw in it a threat to religion, so censorious, as opposed to amused, ridicule of any practice that could be related to religion, and public controversy over any matter of religious delicacy, affronted or disturbed even many intellectual thinkers who genuinely regarded serious satire in religious connections as detracting from the atmosphere of faith and devotion. Their fears were not peculiar to Spaniards. In England, in 1796, for example, the posthumous voice of Gibbon rose prominently among international voices:

> The conclusion of my work [the *History of the Decline and Fall*] was generally read, and variously judged. The style has been exposed to much academical criticism: a religious clamour was revived, and the reproach of indecency has been loudly echoed by the rigid censors of morals. I never could understand the clamour that has been raised against the indecency of my three last volumes . . .[73]

In Spain, Isla told a correspondent about the objections to the publication of the First Part of *Fray Gerundio* by the Bishop of Palencia, from whom he had applied for its licence: objections, so the context suggests, concerning method rather than matter.[74] A scholarly friend, J. M. de Santander, Academician and Royal Librarian, who otherwise approved of *Fray Gerundio* and supported Isla's evidence of outrageous sermons with innumerable examples from Library archives, was still troubled about the publicizing of complaints before the general public.[75] Another eminent friend, Miguel de Medina, warily described *Fray Gerundio* as medicine necessarily 'acre' (bitter).[76] Isla's most illustrious sponsor, Agustín de Montiano y Luyando, found time in his eulogies to mention the ticklish, 'vidrioso' (glassy), nature of Gerundian subject matter, and to anticipate criticism.[77]

Foremost among reasonable objections to the appearance of *Fray Gerundio* was the fact that the satire had been published during Lent, the season of solemn recollection when caricature was out of place. For however much Isla might later excuse himself by saying that the timing was due to his publisher's miscalculation,[78] the damage by then had been done and the charge of scandal spread abroad. Also into Isla's open praise of his own and the Benedictine Order – the notably intellectual Brotherhoods – could be read a tone of superiority, especially in that the word 'Fray', as applied to the caricatured protagonist, was a form of address used most commonly among members of Mendicant Orders – Franciscans, Augustinians, Dominicans and Carmelites – as opposed to the word 'Padre' customarily used in preference to the unassuming 'Fray' by Benedictines and Jesuits. Indeed, beneath the surface, though emerging more maliciously in the banned Second Part, which circulated subterraneously, was a clerical arrogance of intention probably more vicious than Isla himself realized. His reference, in circumstances of odious comparison by implication, to unholy characteristics of Fray Gerundio's Order, which of course was not more precisely named than the word 'fray' suggested, could hardly have been more provocative. He sneered at its mercenary motives in promoting the interests of wealthy and powerful personages while ignoring or condoning their loose living;[79] at alleged dishonesty, underhand manipulation in monastery government, 'cheap' absolutions to beautiful or important women; the hypocrisy of outward docility, for the purpose of personal advantage, to seniors open to flattery; and indecorous behaviour with nuns and women generally.[80] In fact, concluded Isla, such principles of behaviour give support to Protestant critics of the Church in Spain.

These criticisms, justified as they may have been either in general or regarding a particular Order, and amusingly ingenious as was the manner of their expression, bore the stamp of libel. So that many of Isla's opponents reasonably argued that such criticism would more properly and compassionately have been voiced in serious, private discussion among the clergy rather than in a public display of ridicule; and that this was a form of satire which Isla, whose office first and identity afterwards soon became apparent, obviously relished with an all too secular malice. If his religious critics failed to be amused, they could hardly be blamed for mere stupidity. It was natural, too, that, scandalized by the near libel of much of *Fray Gerundio*, they should have allowed themselves to be scandalized by other features of the work which in themselves were neither scandalous nor defamatory.

The secret of Isla's identity under its safeguarding pseudonym hardly survived the first weeks of publication. Doubtless his friends in high quarters and brothers in religion, jubilant at the book's success, were partly responsible for the leakage.[81] Hostile critics operated by underground inquiry. Soon, therefore, manuscript letters and printed pamphlets, with or without licences, began to circulate in protest, and not all the efforts of Isla, his Order, and his influential friends in high places, could soften their effect. *Fray Gerundio* had amused the King and had been accepted by the Pope because they approved of its general theme of reform and were impressed by its ingenuity. But, when it became evident from Isla's particularities that he had offended the Mendicant Orders and secular clergy, no Royal decree was forthcoming, as in Feijoo's experience in 1750, to silence every critic. By 1758 too the Society of Jesus was already under critical, international scrutiny. All that Isla's supporters could do was to try to ensure that criticism against him should be contained within bounds officially licensed.

There was theoretic and practical reason, for the times, in many of the objections brought against Isla by men whose views, we now, from hindsight, should call ignorant. An 'Amador de la Verdad, Carmelita Descalzo' (Lover of the Truth, Discalced Carmelite . . .), *alias* Juan de Chindurza,[82] who put into circulation a manuscript letter to Isla in February 1758, apparently was first to complain about his peculiar method of reforming clerical abuses. And although the Jesuit might simulate righteous indignation at this and other criticisms reaching him, and complain of malicious intention within his critics' own method of attack, he was hardly in a position to defend himself convincingly, if, with false or half false intention, they offered him supposedly friendly advice in honeyed tones of disgust. Isla himself, whose voice in *Fray Gerundio* was anything but pious and whose tones of double dealing were only partially edifying, had been, after all, the first to offend. So the Lover of Truth begins by explaining why the appearance of what looked like a theatrical farce (*entremés*) during Lent, the season when Spanish theatres closed down, gave cause for serious concern. Not, he asserted, perhaps in all honesty, that he was objecting to Isla's complaint about bad sermons and bad preaching techniques. What he deplored, he said, was the Jesuit's unseemly manner, and, in content, his barbed insinuations about monastic life in other Orders than his own. Cheap publicity, he went on, is not the right means of correcting human defects in the Church's representatives. Rather it had been the practice of holy men, however zealous in their desire to eradicate abuses, to report straightforwardly and circumspectly to the proper authorities, because:

> Sus virtudes y su comprensión les hizo creer no eran decentes medios las mojigangas, las chufletas y las ridículas burlas, para corregir a personas sagradas, a las cuales se les debe tratar con modo reverente y corrección secreta, aun en el caso que se reprendan sus abusos; porque la publicidad de sus defectos ocasiona grandes inconvenientes en la Iglesia . . .[83]

> (Their virtues and understanding caused them to believe that masquerades and jests were not decent means of correcting persons whom one should treat with respect, and who should be corrected in private even for the purpose of rebuking their improprieties; because to publicize their defects produces grave problems in the Church.)

So far the Lover of Truth spoke with reasonable objectivity. But, foreseeably, Isla's assumptions of Jesuit superiority had suggested to his critic an all too convenient form of retaliation. How, he then proceeded vindictively, how would the worthy Society of Jesus feel if some critic, professing to call for the reform merely of minor religious defects, such as unintelligent preaching, should go out of his way to refer to the behaviour of a minority of Jesuit missionaries in Paraguay, and to the scandal of their militant activities against the Spanish Crown? This may seem an ugly example of tit-for-tat. Yet if the stage were now set for ugly polemic, it was not the Amador de la Verdad who had prepared the action or was responsible for the disapproval of the Inquisition. Matching Isla's menace with menace, the Amador de la Verdad certainly derived pleasure from believing that the Inquisition was unlikely to be impressed by those explanatory, sponsoring letters prefacing the 1758 edition of *Fray Gerundio*. Indeed, given the inherited responsibilities of the Inquisition, which was revered by the majority of Spaniards, it must be remembered, as protector of the Faith, it would have been almost impossible for Inquisitional officers to have failed to take action. Similar action had already been taken in similar circumstances by foreign protectors of foreign forms of Faith. In supposedly liberal England Gibbon spoke of anti-Catholic prejudice actively directed to the ends of safeguarding Protestantism; of the alarm raised in the University of Oxford at reports about Catholic missionaries there:

> By the keen Protestants, who would gladly retaliate the example of persecution, a clamour is raised of the increase of popery: and they are always loud to declaim against the toleration of priests and Jesuits who pervert so many of his Majesty's subjects from their religion and allegiance. On the present occasion, the fall of one or more of her sons directed this clamour against the University . . . ;

of his own expulsion from Magdalen College after the 'high treason' of his conversion; of religious and moral opposition to his *Decline and Fall*, especially on grounds of his supposed 'indecency' in painting the manners of Roman times.[84] France supplies similar examples.[85] Thus, by the moral codes of the period, offences against the established Defenders of the Faith in any country, Catholic, Anglican or Protestant, were commonly considered equivalent to treasonable offences against the State, and could logically incur similar penalties. In condemning the Spanish Inquisition out of context, critics have tended to forget its European equivalents, and its accepted rôle as more protective than inimical.

In no country was the public mind so rational that a Swift, Gibbon, Descartes, Voltaire, Montesquieu, Martínez, Feijoo or Isla could render it unshockable. Certainly both Dean Swift and Father Isla might be said to have gone too far in personal malice for any age to find their professed intentions unobjectionable. So the fundamental question seriously presented by the Amador de la Verdad was, beneath its polemical trappings, timelessly realistic. Is it seemly, legitimate, or Christian, for a member of an Order, or for any ordained person, to direct satire for religious purposes against other clergy or religious? It was a question consistently occupying the minds of most of Isla's critics. The moral, philosophical or theological evidence which some of them brought to deny Isla the right to ape Cervantes in a sacred context, was sometimes irrational and ill-informed to the point of fatuousness: a matter which the Jesuit took care to dramatize. Yet he never satisfactorily answered their basic objection to his disregard for clerical decorum. Where artistic instinct and spiritual humility came into conflict in Isla's make-up, artistic instinct tended to win.

The same serious question was given prominence, also in 1758, by a so-called 'Penitente de Marquina' (One of Marquina's Penitents) who, according to Isla, was the Capuchin Friar Matías Marquina himself, in a paper described by the Jesuit as 'crazed'.[86] Justifiable as that adjective might be in certain respects, the Penitente's 'craziness' does not affect the sense of his general argument, which is that Isla had transgressed against clerical propriety by libellously sniping against the regular clergy and certain Orders. Nor are all his minor arguments unworthy of attention. He admitted, as did most honest clerics, that preachers had tended to exaggerate the importance of rhetoric and to subordinate Christian fact to popular fantasy. His classically refined ideal of rhetoric, it will be remembered from his *Escuela general histórica . . .*, had been akin to that of Isla. Where the Penitente becomes crazily pedantic is in the unbalanced Scholasticism of his incidental questioning of Isla's supposed evidence. Do Isla and his sponsors assume, he asks, that satire for religious purposes is good *per accidens* rather than *per se*?; that, if the Fathers had used this weapon, they might have corrected abuses current in their times? But to assume as Isla must, therefore, that the Church Fathers failed to adopt right methods of teaching, is blatant blasphemy:

> . . . hiere a la majestad de Cristo, nuestro Señor, con herética blasfemia . . .[87]
>
> (. . . it strikes at the majesty of Christ, our Lord, with heretical blasphemy . . .)

To the mind of the Penitente, *Fray Gerundio* had caused a scandal in the Orders, amusement among the licentious, and a state of preparation for heresies.

Isla was right to classify such pedantic extremism as a logician's raving and to dismiss his '*ergo* y más *ergo*' (*therefore* and more *therefores*) with a trenchancy worthy of Feijoo.[88] Yet embedded in the Penitente's raving are more astute observations whose effect is spoilt by the author's manner of argument. One concerns the effect of *Fray Gerundio*'s publicity on ordinary readers who prefer sensational colour to the starkness of basic truths; on foreign readers, too, who use editions printed without explanatory Prefaces and who therefore are liable to receive wrong impressions of the Church in Spain. As the Amador de la Verdad would have agreed, the general public is given to generalizing outwith general contexts. Isla's want of fraternal charity, however amusing artistically, was another feature of *Fray Gerundio* reasonably criticized by the Penitente as improper in a religious. But more interesting to a modern reader is the Penitente's acumen in querying Isla's aim. Was his purpose really that of a reformer, or that of an entertainer? The Penitente in his 'crazy' simplicity had touched the pulse of a psychological paradox: the reason for the Jesuit's unreasonable manners. Father Isla, we come to realize as we read *Fray Gerundio* and his self-conscious writing elsewhere, was by instinct primarily an artist and a reformer by afterthought; was ambitious enough to see himself as an associate of international satirists competing for intellectual attention on the plane of rationalistic wit; to value himself not merely as a Swift or Montesquieu but as an eighteenth-century Cervantes or Quevedo; was, by his Quevedan and Torresian lineage, conversant with the potentialities of familiar ironies, those provocatively within range of his professional experience.

Most telling of all the Penitente's saner observations is his remark that oratorical excesses in the pulpit are the result partly of sensational demand in congregations. Which brings us back to the emotional needs of a period of mental disturbance halfway between Baroque puzzlement and scientific rationalism. While philosophers and literary thinkers logically reasoned, the masses, though not untouched by change, responded imaginatively, in their need for authoritative pronouncement, to the orchestral warmth of elaborate oratory, reassured by emphatic sound with which popular drama had familiarized them, soothed by the very fact that the crowded imagery, so confidently presented, sounded transcendently confident. These sounds corresponded to the stimulating harangues of assertively strong characters in the theatre where a general effect of sound-power was more important to the *vulgo* than every detail of meaning. In Church, as in the theatre, a general religious fervour was commonly created by decorative histrionics and the personality of the speaker. In times deficient in sensational entertainment the declaiming preacher must have supplied emotional therapy. Plain speaking in ordinary tones about ordinary circumstance and common conscience could have no extraordinary appeal to average hearers deprived of new sources of excitement. Sound, as a leading English actor and theatre-manager of the times, Colley Cibber, explained from experience, has a histrionic value in its own right, and he relates the intrinsic potentiality of the spoken sound to that of music:

> If what was truth only could have been applauded, how many noisy actors had shook their plumes with shame, who, from the injudicious approbation of the multitude, have bawled and strutted in the place of Merit? . . . So that there is even a kind of language in agreeable sounds, which, like the aspect of beauty, without words, speaks and plays with the imagination.[89]

How often did the conscientious John Wesley mistake the temporary hysteria of his hearers for permanent conviction? Perhaps it was a folk-necessity. Perhaps displays of unanalytical emotionalism still are, in many quarters, folk-necessities. Certainly the time was approaching in the Age of Reason when, for intellectual purposes and industrial development, not even the elaborately simulated plainness of Torres or Isla, but unequivocal rationalism would be substituted for oratorical sensationalism. In the meantime the nervous sense of change was cushioned for average congregations by a padding of rich verbiage, and for theatre-audiences by operatic insinuation and the atmospheric effects of new stage-machinery.

Isla, in private letters and in *Fray Gerundio*, together with some of his supporters and critics, refers to the custom whereby invitations to preach on special occasions, during Lent, or for general or local festivals, were commonly delivered at the instigation of a local dignatory or a lowly *alcalde* (mayor): that is, often enough, persons and groups, not themselves intellectually critical, and disposed to judge a preacher by his popular appeal.[90] The Penitente's reminder to Isla that much of the blame for bad preaching lay in the popularity of bad preachers is emphasized in his chapter of conclusions where he speaks of the 'sobra de ignorancia en los oyentes' (superfluity of ignorance in the hearers).[91] The fact that congregations 'gustaban de los sermones que no entendían' (liked the sermons which they did not understand) and were disillusioned by simple sermons which any child could understand,[92] may be difficult for a modern reader to appreciate. Yet it was a fact as naturally true of Church congregations as it was of theatre-audiences. The masses, Spanish or foreign, were used to general impressions of sound and visual effect, were ready, when listening to a sermon, to resort to their own imagination, guided by atmospheric pointers, just as they resorted to their own imagination – with or without theological sanction – during the course of a Mass or an Anglican Office. It was the scientific rationalism of the Age of Enlightenment that enforced the necessity of literal understanding in mass gatherings, and neither Spain nor any other country learned that new technique quickly. The Penitente, therefore, was right to remind Isla of the public demand, of which he too disappproved, that a preacher should speak 'en cadencia' – that is, speak in sonorous splendour even if with 'hollow words'.[93] Not least significant in this respect is his reference to the preachers' habit of using as titles of their sermons the titles of popular plays, especially, one might add, titles concocted in the typically paradoxical wording of seventeenth-century dramas that combine an overall suggestion of complicated ideas with a grandeur of sound: Alarcón's *No hay bien que por mal no venga* (*There Is No Good That Has Ignored Evil*); Matías de los Reyes' *Di mentira y sacarás verdad* (*Say It's a Lie and You Will Extract Truth*); Calderón's *Sueños hay que verdades son* (*There Are Dreams That Are Truths*); or *En esta vida todo es verdad y todo mentira* (*In This Life All Is Truth and All Is Untruth*), and the like. The practice was confirmed by, among others, one of Isla's sponsors, a Royal Chaplain, José de Rada y Aguirre, who is satirized by Isla in *Fray Gerundio*,[94] and recently has been recorded by Aguilar Piñal, who, in *Impresos sevillanos del siglo XVIII*, lists religious works employing titles of dramas, including the funeral oration on the death of Felipe V, which declared, in the title-words of Vélez de Guevara's drama, that the King 'reina después de morir' (reigns after dying).[95]

In literary practice, of course, neither the Lover of Truth nor Marquina's Penitente was a match for the intellect and artistry of Isla or for the polemical proficiency of his defenders. The series of Letters published in 1787 answering objections raised by the Penitente were devastatingly rational.[96] His pliable logic was as clay in the hands of his Jesuit manipulator, who jeered at his unsteady Scholastic pretensions:

> Amigo mío digo que digo que cuando digo digo, no digo digo, sino digo que digo que no digo.[97]
>
> (My friend I say that I say that when I say I say, I do not say I say, but I say that I say that I do not say.)

Isla could point to mistakes in the Penitente's translation from Latin as well as to certain incorrect references to the Classics and Scriptures.[98] He attacked his confusing of satire with libel.[99] The Penitente's pseudo-Scholastic twist by which Isla was accused of blasphemy made the Jesuit, so he said, roar with laughter,[100] and was neatly turned, in retaliation, into the contention that the blasphemer was the Penitente himself for saying that Christ did everything he 'could' to put the world to rights, so assuming His power to be limited.[101] As for objections to the use of the word 'frailes' (friars) as representatives of religious defects and abuses – if that cap fits, implies Isla with satisfaction, let the Penitente wear it. Nor, emphatically pursued the supporting voice of Don Cernadas,[102] nor do Protestants and other heretics read into Catholic criticism of Catholic defects a support of Protestant doctrine:

> . . . los herejes, aunque pierden la fe, no pierden el uso de la razón.[103]
>
> (. . . the heretics, though losing their faith, do not lose the use of reason.)

Though here Cernadas was not allowing for the fact that examples of Catholic error distributed publicly abroad might well provide useful illustrations of Catholic error to a Protestant *vulgo* which, like any other *vulgo* then, or now, would use them out of context. As for his critics' condemnation of unnecessarily dirty references in *Fray Gerundio* – to excrements, the letting down of trousers etc. – Isla supplied some indelicate references in one of Marquina's sermons to a beautiful birthmark on a lady's breast where, he had said, she had hidden it that it might be reserved for the gaze of her Beloved.[104] Such an exchange of accusations advanced the moral cause of neither party and contributed to the general atmosphere of scandal.

It was all very well and right for Isla and his supporters to protest that standards in theatres and churches ought not to be set by groundling congregations. It might have seemed then, and would still seem reasonable, for him to argue that the masses are more likely to pay attention to pleas for reform where pleas take an entertainingly novelesque shape, as in *Fray Gerundio*, than to pedagogical lectures. But the enlightened Isla and his supporters oversimplified a very complicated state of mind in the general public. The Penitente himself, rather than defending oratorical excesses, was expounding, however badly, a psychological problem concerning public taste, for which *Fray Gerundio* offered only a negative solution. Change would depend on changing experience. It would come slowly, and then only partially, with the experience in everyday life of the industrial need to observe scientific cause and effect, and the consequent need to revise educational techniques. Meanwhile reformers had to learn to accommodate themselves to scandalized opposition that stemmed less from stupidity than from logical habits of thought and feeling and from pre-rationalistic associations of ideas.

Even so, even when every allowance is made for Isla's best religious, moral, and academic intentions, there remain as objects of moral reproach his malicious tone, his snide insinuations, and odious comparisons by implication, that, from any clergyman, would always scandalize a majority of readers. The Father Master Feijoo had been brutally blunt, not calculatedly malicious. He had a considerable conceit of himself. But his dominating self-confidence derived from his faith in his cause and in his constructed evidence, not primarily from personal showmanship. Cervantes, with whom Isla was always comparing himself, was a philosopher-satirist whose objectivity was above Isla's experience and whose many-layered irony was beyond the voice-range of any eighteenth-century reformer. Much nearer to Isla's understanding was the waspish invective of Quevedo, whom he regarded as a master. For which reason Isla is at his most comprehensible in the uncomfortable company of the virtuoso Torres Villarroel, with whom he was on the worst of reformist terms.[105] Of such apparently unlikely stuff is the history of ideological polemic made, and through such labyrinths of popular assumptions must reformers learn to move.

Whatever Isla's real or pretended motives may have been in making his Gerundio a friar, however well justified he may have seen his rôle as a Catholic publicizer of Catholic abuses, and however expertly he and his defenders explained themselves, the reproach of unseemliness persisted. A few months after the appearance of *Fray Gerundio*, a Padre Don Juan de Arabaca, who described himself as a 'Misionero del Oratorio del Salvador' (Missionary of the Oratory of the Saviour), disgustedly summarized the kind of objections being made to Isla's method of reform.[106] Controversy had broken out over Isla's alleged plagiarism of the theologian Barbadiño, and Arabaca's letter is intended as a reply to the most distinguished of Isla's sponsors, Agustín de Montiano, who had forwarded to him a *Carta escrita por el Barbero de Corpo* written in Isla's defence.[107] Arabaca, who says that he had refused to act as one of Isla's sponsors, reveals his own anti-Jesuit prejudice when he begins by commenting sarcastically on the Jesuit assumption that

> . . .es delito irremisible el no conformarse en todo y por todo y a ojos cerrados con cuanto dicen y quieren los Jesuitas.[108]
>
> (. . . it's an unpardonable crime to fail to conform blindly in everything and regarding everything with everything said and required by the Jesuits.)

He professed, for what his profession was worth, not to believe that Isla had answered the reproach of plagiarism convincingly. But he then passed to a consideration of what so many regarded as Isla's chief offence against fraternal charity and religious seemliness, incidentally disposing of the Jesuit's self-comparison with Cervantes, which, in times of Jesuit unpopularity, seemed a good sample of Jesuit arrogance. The passage concerned is redolent of strong feeling more human than academic:

> . . . que se parece al Quijote de Cervantes, como las coplas de Benegasi a las églogas de Garcilaso.[109]
>
> (. . . he is as comparable to Cervantes' Quijote, as are the verses of Benegasi to the eclogues of Garcilaso.)

Similarly a 'Romance contra Fray Gerundio escrito por el Marqués de la Olmeda', while recognizing Isla's 'great talent', advises him to apply his genius to writing novels of worldly, instead of religious, subjects, calls his attention to the sobriety and charity expected of members of the Orders, and, protesting that Isla's theme is too delicate to be treated satirically, describes *Fray Gerundio* as the

> escándalo de los claustros
> y alboroto de las celdas.[110]
>
> (scandal of the cloisters
> and disturber of cells.)

Another set of verses, 'Endechas del Padre Marco', which refer to Isla's 'grossness', declares that the satirist's own sermons have been known to transgress in ways condemned by Isla in other clergy:[111] a point duly denied in a defensive *Diálogo entre el cura del Zángano y el guardián de Loriana.*[112] Within a full chorus of disapproval certain voices insert more caustic notes of disgust over Isla's grossness, and seize the opportunity to interpret his laughter in spiritual matters as Jesuit conceit. The verses 'Décimas de un cocinero de cierta religión contra Fray Gerundio' are shrill with retaliatory spitefulness.[113]

Perhaps the atmosphere generated by *Fray Gerundio* has much in common with the atmosphere engendered by Swift's *Tale of a Tub* which, when published in 1704, when Swift was already ordained, shocked English bishops into protesting to Queen Anne about the exhibitionist ribaldry indulged in by satirists of religious abuses. Malice, if artistically excusable, could be socially embarrassing. And the printed circulation of artistic malice, like the public progress of the word 'scepticism', provided a serious cause of misunderstanding.

As was already apparent in certain of the licensing *censuras* of *Fray Gerundio*, even Isla's supporters, who approved of the book as a whole, doubted the wisdom of some of its details. A pamphlet in the form of a series of Letters, *Cartas aldeanas*, written in praise of the work, so its authors say,[114] pinpoints criticizable features by conspicuously defending them: the delicate matter of the use of the word 'fraile', the moral justification of religious satire, together with various 'lunarcitos' (little beauty spots) of anachronism, occasional pedantry and long-windedness.[115] In explanation of blemishes the reader is referred, possibly with conscious inadequacy on the author's part, to Isla's explanatory Preface. Most telling, however, is a calculatedly diffident admission that, if *Fray Gerundio* has its defects, the author's 'indiferencia y libertad histórica' (indifference and historical liberty) – which in the context means Jesuit lack of disinterestedness – may be counted as one of them.[116]

When all arguments of all contenders are gathered together, the overall impression they give is of unease: an unease which indeed constitutes an early stage in the life-process of change. As in experimental sciences, so, in oratory, the average man was being asked to distrust the evidence of his senses, the sources of his psychological security, the colour and companionship of custom, even some of the objects of his inherited, legitimate respect and worship. In time, place and condition the ideological crisis developed among the many with a complexity natural to average reasonableness. What we now find difficult to understand is that both the fully confirmed rationalists and the fully entrenched obscurants were few, that both extreme groups had to contend not simply with each other, but with intangibly bewildered clouds of well-meaning witnesses to apparent reasonableness. Simple intent was of necessity diversified into elaborate apologetics. The social truth of the mental times, therefore, lay neither in philosophical assertions nor denials, but in the nervous atmosphere which they engendered.

NOTES

1 *An Apology for the Life of Mr Colley Cibber*. Everyman, No. 668 (London, 1938), p. 54.

2 See p. 98 below.

3 See León Lopetegui, 'Visión de la oratoria sagrada de dos destacados escritores del siglo XVIII: Mayáns y Siscar-Burriel (1744-1758)' in *Letras de Deusto* (Bilbao), VII (1977), No. 13, p. 89.

4 León Lopetegui, *art.cit.*, p. 105, and see Feijoo, 'La elocuencia es naturaleza y no arte', *Cartas eruditas*. C.C., *ed.cit.*, pp. 37ff.

5 *Fray Gerundio, ed.cit.*, Vol. II, p. 189, *et passim.*

6 *Las cinco piedras de la honda de David, en cinco Discursos Morales, predicados en Roma a la Serenísima Reina de Suecia, Christina Alexandra, en lengua italiana, y traducida en castellano por el mismo autor* (Madrid, 1711). Vieira is discussed by Isla's characters in *Fray Gerundio, ed.cit.*, Vol. II, pp. 190ff.

7 Vieira, *op.cit.*, pp. 10-12.

8 *Op.cit.*, p. 35.

9 José Maimó y Ribes, *Defensa del Barbadiño.* I have seen a copy in the Biblioteca Nacional, Madrid, without author's name, title or year of publication, pp. 97-98. See p. 94.

10 *Carta pastoral . . . en que manifiesta a todos sus súbditos los motivos que hay para temer que la ignorancia de las verdades cristianas es mayor de lo que se hace juicio, para que todos, en cuanto les sea posible, soliciten el remedio.* Bib. Nac., n.p., n.d., p. 163. See also, for example, Francisco Joseph Artiga, *Epítome de la elocuencia española*, 2nd edition (Pamplona, 1726); the section 'Naturaleza de la oratoria sagrada' in *Resurrección del Diario de Madrid . . .* (Madrid, 1748), etc. A useful article on the century's view of eloquence is, for example, Don Paul Abbott, 'Antonio de Capmany and the New Rhetoric', *Dieciocho*, VII (1984), No. 2, pp. 146 ff.

11 Vieira, *op.cit.*, p. 174.

12 J.B. Masillon, *Œuvres complètes*, ed. E.A. Blampignan (Bar-le-Duc, 1865).

See *Sermon pour la fête de Tous les Saints* . . . preached in 1699 before the King and Court, pp. 112 ff. Compare a comment on French preaching by P.G. Marivaux: '. . . c'est là le vice de nombre de prédicateurs; c'est bien moins pour notre instruction qu'en faveur de leur orgueil qu'ils prêchent . . .'. *La Vie de Marianne*, Part IV, 1736, ed. M. Gilot (Garnier-Flammarion, 1978), p. 199.

13 See Approbation of Fr. Franc. de Madrid, *Oración fúnebre panegírica* . . . (Madrid, 1744).

14 I have seen this 'Nouvelle Édition où l'on ajoint les Remarques de Mr Lenfant' (Amsterdam, 1728).

15 See, for example, p. 383 *et passim.*

16 *Fray Gerundio, ed.cit.*, Vol. III, pp. 117ff.

17 Gisbert, *op.cit.*, pp. 336-37.

18 *Op.cit.*, p. 4.

19 See *Christian Eloquence in Theory and Practice* (London, 1718), Preface.

20 *Op.cit.*, p. 3.

21 See pp. 121ff. below.

22 See the Letter dated Oviedo, Oct. 29th, 1749. B. Nac. MS. 10579. He was referring, for instance, to Francisco de Soto y Marne's *Florilogio sacro* . . . of 1738, discussed pp. 102-03 below.

23 See, for example, Mayáns, *op.cit.*, Vicent Peset, *Gregori Mayáns i la cultura de la il·lustració*, p. 399, *et passim*, and León Lopetegui, *art.cit.*, pp. 105-06.

24 See Agustín Hevia Ballina, 'Hacia una reconstrucción de la librería particular del P. Feijoo' in *Fray Benito Jerónimo Feijoo, Studium Ovetense* (Oviedo, 1976), pp. 154-55. See also B.J. Feijoo, 'Glorias de España', *Teatro crítico universal*, Clásicos Castellanos, Vol. II, pp. 191 ff; and Feijoo's letter to Mayáns on the subject: Ant. Mestre Sanchis, *Ilustración y reforma de la Iglesia. Pensamiento político-religioso de Don Gregorio Mayáns y Siscar* (Valencia, 1968), pp. 91-92, n. 8.

25 See, for example, many of the pamphlets and letters printed in BAE, XV.

26 Isla, *Sermones morales* (Madrid, 1792), Vol. III, pp. 1 ff.

27 *Op.cit.*, Vol. I, p. 6.

28 See León Lopetegui, *art.cit.*, pp. 95 ff.

29 Mayáns, *El orador cristiano* (Valencia, 1733). See pp. 50 ff, *et passim.*

30 *Op.cit.*, p. xxviii.

31 *Op.cit.*, Mayáns, as Mestre Sanchis explained (*op.cit.*), was an advocate of interpretation from original sources and therefore of the training of clergy in Hebrew and Greek to which the Jesuits then gave insufficient attention.

32 See León Lopetegui, *art.cit.*, pp. 105-06.

33 *Escuela general histórica, crítica, política, moral, divididas en nueve lecciones sacadas de la vida y virtudes del Glorioso San Antonio de Padua* . . . (Madrid, 1751).

34 *Op.cit.*, Vol. I, pp. 11-12.

35 *Op.cit.*, p. 175.

36 *Op.cit.*, p. 192.

37 See the 'Memorial y manifiesto que al Rey Católico de las Españas, y dos Sicilias, El Sr. Don Carlos de Borbón, presentó Don Alfonso Carlos de Ribera . . . natural del Reino de Galicia'. B.Nac. MS. p. 156.

38 See, for example, Letters CXXIII, CXXXIV, *et passim*, *Cartas familiares*, BAE, XV, Part I, etc.

39 See, for example, *Fray Gerundio*, *ed.cit.*, Vol. II, Chapters V and VI, *et passim.*

40 *Op.cit.*, Vol. II, p. 119.

41 *Defensa del Barbadiño*, p. 32. The copy I have seen in the Biblioteca Nacional, Madrid, bears no title page. Maimó y Ribes' criticism is contested in its turn, though unconvincingly, by 'el Barbero de Corpa' in a *Carta escrita por el Barbero . . . a Don*

José Maimó y Ribes (1758). See BAE, XV, pp. 359 ff. The 'Barbero' is evidently one of Isla's pseudonyms.

42 See Isla's papers directed against Feijoo's critics, especially Torres and Pedro de Aquenza, *Colección de papeles crítico-apologéticos, que en su juventud escribió el P. Joseph Francisco de Isla . . .*, Parts I, II (Madrid, 1788).

43 *Fray Gerundio, ed.cit.*, Vol. IV, p. 125.

44 *Op.cit.*, Vol. IV, p. 267.

45 See, for example, *Fray Gerundio, ed.cit.*, Vol. I, pp. 105-06, 109-110, Vol. III, p. 46 etc; *Cartas de Juan de la Encina*, BAE, XV, p. 404; *Cartas familiares, op.cit.*, p. 629 (*Carta* XXXVIII), etc.

46 *Sermones morales* (Madrid, 1792), 5 vols. Vol. I, p. 155. The extract is from a sermon preached in 1733.

47 See, for example, his letter CXXXIV, BAE, XV, p. 474.

48 *Cartas.* BAE, XV, pp. 597-600. See p. 598.

49 See *El orador cristiano* (Valencia, 1733).

50 See the article by N. Rochaix, 'José Climent et la lutte contre l'ignorance dans l'Espagne du XVIII Siècle', *Les Langues Néo-Latines*, LXXII (1978), No. 1, 26-64. See p. 61. Also see *Fray Gerundio, ed.cit.*, Vol. II, p. 230, referring to the reformer Francisco de Valero y Losa.

51 Swift, *Tale of a Tub* (with *Gulliver's Travels*) (London: Hamish Hamilton, 1930) (The Modern Library), pp. 460, 463, 478, 497.

52 The *Spectator*, London, Everyman, Vol. I, No. 62 (May 11th, 1711); No. 147 (Aug. 18th, 1711).

53 See Wesley, *Sermons* (London, 1825), Vol. II, p. 68.

54 *Op.cit.*, Vol. II, pp. 35 ff.

55 See *Journal*, Everyman, 1938. Vol. II, pp. 452-53, p. 490, *et passim.*

56 *The Life of Samuel Johnson, ed.cit.*, Vol. I, pp. 390-91.

57 Voltaire, 'Mais il y a d'anciennes bornes qu'on ne remue pas sans de violentes secousses', *Siècle de Louis XIV* (Amsterdam, 1774), 2 vols. See Vol. II, Chapter XXXV, p. 459.

58 See Act II, xi.

59 See *El perro y la calentura. Novela peregrina.* 'Por Don Francisco de Quevedo, quien la imprimió bajo del [*sic*] nombre de Pedro Espinosa' (2nd edition, Madrid, 1736). It is referred to by a 'Don N. Cernados' in a 'Carta del Lirondo Crisanto Criton'. See BAE, XV, p. 288.

60 For examples of all these objects of ridicule see *Fray Gerundio, ed.cit.*, Vol. II, p. 145, 155, 208; Vol. III, pp. 20 ff; Vol. IV, pp. 139, 200 ff, 267-68, etc. See also the allegedly authentic examples given in the 'Carta del Lirondo Crisanto Criton', BAE, XV, pp. 282 ff, *et passim.*

61 Feijoo, *Teatro crítico universal*, Vol. VII, no. 8. See BAE, LVI, pp. 381 ff. ('El Toro de San Marcos').

62 Isla, *Fray Gerundio, ed.cit.*, Vol. III, pp. 136 ff and *Sermones morales, ed.cit.*, Vol. I, p. 324.

63 *Florilogio sacro . . .* (Salamanca, 1738), Prólogo. There is no pagination.

64 *Op.cit.*, p. 68. See, for example, Sermons I, II, and III on St Andrew, the Virgin Mary and St Francis.

65 *Sermón en la solemnísima translación del Sto. Sacramento del antiguo al nuevo templo de la Sra del Pilar* (Zaragoza, 1719), p. 14.

66 Fr. Juan Rumualdo Agramonte, *Sermón que compuesto de paraenético y panegýrico, la Sagrada Religión de Predicadores, a un tiempo siente y celebra la muerte y la inmortalidad del Rev.ísimo P.Fr. Juan de Soto.* (Alcalá: Joseph Espartosa, 1736).

67 P.Fr. Alexandro de la Concepción, *Oración continua, y retórica que en acción de gracias a la Beatísima Trinidad, por la felicísima redempción, últimamente ejecutada en la ciudad de Argel por los Padres Trinitarios* . . . (Madrid: Ant. Sanz, 1755).

68 P.Fr. Alonso de Huecas, *Oración encomiástica, de las admirables virtudes que con extraordinario medio, y discreción sutil se predicó el dia 4 de octubre de 1756* . . . (Madrid, 1757), p. 3.

69 *Op.cit.*, p. 1.

70 It was published in Zaragoza, 1704.

71 P.Fr. Alonso de la Guardia, *Oración fúnebre panegírica* . . . (preached on Nov. 6th, 1743), Madrid, 1744. See ante-Introduction, pp. 1-7.

72 Ant. Mestre Sanchis, *Historia de la Iglesia en España*. Vol. IV., ed. Ricardo García Villoslada (Madrid: Editorial Católica, 1979), pp. 604-06.

73 E. Gibbon, *Autobiography*, Everyman, No. 511 (London, 1948), p. 173, first published two years after his death.

74 See BAE, XV, *Carta* CV, p. 463.

75 See BAE, XV, pp. 44 ff.

76 See his letter replying to Isla's requests for his opinion of the work, BAE, XV, pp. 53-54.

77 See BAE, XV, pp. 36 ff.

78 See *Carta* CXXIII, BAE, XV, p. 469, and Isla's polemical *Carta* to Marquina, BAE, XV, p. 344, and pp. 109ff below. For further discussion of Isla's rôle in polemic see A. Labandeira Fernández, 'En torno a la polémica del *Fray Gerundio*', in *II Simposio sobre el padre Feijoo y su siglo* (Oviedo: Cátedra Feijoo, 1981), Vol. I, pp. 111-21.

79 *Fray Gerundio, ed.cit.* See, for example, Vol. II, pp. 142-43; Vol. III, pp. 6, 53; Vol. IV, pp. 17-18, 111 ff, 194, etc.

80 *Op.cit.*, Vol. II, pp. 178 ff; Vol. III, pp. 55-56, 93; Vol. IV, pp. 9 ff, 180, etc.

81 He is named directly or by implication in protests reproduced in BAE, XV, pp. 297 ff, 359 ff, 365 ff, 394, 398. See also León Lopetegui, *art.cit.*, pp. 106-10 which refers to early knowledge of 'Lobon's' identity in the Mayáns y Siscar-Burriel correspondence.

82 His identity has been given as Don Juan de Chindurza of the Secretaría de Estado. See BAE, XV, pp. 259 ff, and Rafael Olaechea, *Política eclesiástica del gobierno de Fernando VI, La época de Fernando VI* (Oviedo: Cátedra Feijoo, 1981), pp. 222-23.

83 Amador de la Verdad, see BAE, XV, p. 260. See also the manuscript letters by Chindurza in the manuscript collection of the Biblioteca Nacional, Madrid. No. 7215.

84 Gibbon, *Autobiography, ed.cit.*, pp. 53 ff, 173, etc.

85 See, for instance, the examples recorded by Voltaire in the final section of his *Siècle de Louis XIV, ed.cit., passim.*

86 See *Reparos de un penitente del Padre Fray Matías Marquina, dirigidos al autor de la 'Historia de Fray Gerundio de Campazas'*, BAE, XV, pp. 261 ff; also Isla's letters to his brother-in-law, BAE, XV, Nos. CXXXVI and CXLVI, pp. 475, 484.

87 *Reparos de un penitente* . . ., *ed.cit.*, p. 265.

88 Isla replied to Marquina in a series of sarcastic Letters published as *Cartas apologéticas*. See his *Carta de aquel mismo Quidam para aquel propio Cuidam*, BAE, XV, pp. 319 ff.

89 *Apology for his Life, ed.cit.*, pp. 62-63.

90 See *Reparos de un penitente*, BAE, XV, pp. 269 ff. For an obvious example of forceful, popular appeal see the *Oración panegýrica* . . . of the Carmelite P.Fr.

Agustín de Jesús María, in praise of St Teresa of Ávila (Madrid: Joseph Rodríguez, 1721). (B.Nac. V.E.C.a 532.11). It rises to its climax with the patriotic apostrophe: 'Ahora es tiempo, Castellana Amazona; ahora es tiempo, / hermana de Marte, y Española Teresa, ahora es tiempo . . .' (Now it is time, Castilian Amazon; now it is time, / sister of Mars, and Spanish Teresa, now it is time).

91 *Op.cit.*, p. 269.

92 *Op.cit.*, pp. 269-70.

93 *Op.cit.*, p. 270.

94 *Fray Gerundio*, *ed.cit.*, e.g. Vol. II, p. 226. Other influences of drama are satirized by Isla in *Fray Gerundio*, *ed.cit.*, Vol. II, p. 42; Vol III, p. 8; Vol. IV, p. 238, etc. See also José de Rada y Aguirre, *Carta*, BAE, XV, p. 42.

95 See Aguilar Piñal, *Impresos sevillanos del siglo XVIII* (Madrid: C.S.I.C., 1974), p. 128, *et passim*.

96 See *Cartas apologéticas* . . ., BAE, XV, pp. 309 ff.

97 Isla, *Carta* III, BAE, XV, p. 343.

98 *Carta* I, *ed.cit.*, pp. 312-13.

99 *Carta* II, *ed.cit.*, p. 324.

100 *Carta* I, *ed.cit.*, p. 309.

101 *Carta* II, *ed.cit.*, p. 326.

102 *Apología* . . ., BAE, XV, pp. 271 ff.

103 *Op.cit.*, 'Carta del Lirondo Crisanto Criton', p. 272.

104 See *Cartas apologéticas* . . . *Carta* I, BAE, XV, p. 317.

105 See McClelland, *Diego de Torres Villarroel*, pp. 44, 136.

106 See Juan de Arabaca, *Carta*, BAE, XV, pp. 365-66.

107 It was a defence specifically against the accusation of plagiarism made by Maimó y Ribes for whom Arabaca had acted as sponsor.

108 *Carta*, *ed.cit.*, p. 365.

109 *Op.cit.*, p. 365.

110 'Romance . . .'. See BAE, XV, p. 393.

111 'Endechas . . .' See BAE, XV, pp. 396-97.

112 *Diálogo* . . ., BAE, XV, pp. 297 ff. See p. 299.

113 BAE, XV, p. 398. For verses in favour of *Fray Gerundio*, see BAE, XV, pp. 95-96 and 398 ff.

114 Isla gives the authors as the Conde de Peñaflorida and 'otros dos caballeros de Azcoitia' (two other gentlemen of Azcoitia). See *Carta* CLXXIX, BAE, XV, p. 492. The pamphlet itself is printed in BAE, XV, pp. 367 ff.

115 *Los aldeanos críticos* . . . BAE, XV, pp. 371 ff.

116 *op. cit.*, p. 372.

CHAPTER 5

Witness of the Popular Stage

The fact that the average preacher, popular orator, or popular scribbler, during at least the first half of the eighteenth century, sought to inflame the interest of a cross-class *vulgo* by relying on *corral-* (courtyard theatre-) affectations of speech and tone, gives psychological significance to the type of drama being produced at that period. Specifically, it calls attention to mental and emotional assumptions in his public which the popular dramatist could take for granted, and to which successful actors related their vocal effects. Popular public speakers in any period have tended to mimic the techniques of popular entertainers, and ultimately, therefore, to be indebted to the entertainers' scripts. In times of literary decadence such debts become more noticeable, and during Spain's descent from her Golden-Age peaks they are bleakly conspicuous.

The reciprocal influence of drama on audiences, and audiences on drama, as paralleled nowadays in television entertainment, is indisputable, and apparently operates in general either to their mutual intellectual improvement or mutual deterioration. But in a period of relative decadence, intellectual and artistic, the lack of some dominating stimulus, emanating from the individuality of a creative genius, means that both authors and audiences are left to stagnate on plains of empty emphasis and unexacting repetition, rather than become alerted disturbingly to potentialities of innovation.

None better than Golden-Age playwrights, some of whom had served their apprenticeship on the boards, knew the practical art of phrase and sentence-building that permitted an actor sensationally to raise or drop his voice at strategic moments, to suggest the reserve of tension or the exuberance of relief. Their ingeniously parallel constructions, conveying melodious contradictions of paradox, invited a teasing interplay of overtones and undertones. Their alliterative flourishes created opportunities of vocal beguilement. The voice of the Golden-Age actor would memorably convey memorable ideas by the grace of the writer's technical perception. Literary sense memorably consorted with memorable histrionics.

The celebrated forms, the 'tunes' of famous Golden-Age passages such as the following, strengthened the standardization of declamatory voice-patterns in acting. Here is a Calderonian model for breaking the cumulative insistence of one sound by a sudden, whimsical change of pitch, at the end of the section:

Valiente moro y galán,
si adoras como refieres,
si idolatras como dices,
si amas como encareces,
si celas como suspiras,
y si como sientes amas,
dichosamente padeces. (*El príncipe constante,* Act I)

(Valiant and gallant Moor,
if you adore as you relate
if you idolize as you say [you do],
if you love as you insist [you do].
if you are guarded when you sigh,
and if you love as you feel,
[then] you suffer fortunately.) (*The Constant Prince*)

Another Calderonian practice-model offers a variety of shout-inflexions, or, to the creative actor, a complex tune-pattern of shouts, whispers and insinuations:

¿Qué es la vida? Un frenesi.
¿Qué es la vida? Una ilusión,
una sombra, una ficción,
y el mayor bien es pequeño;
que toda la vida es sueño,
y los sueños sueño son. (*La vida es sueño*, Act II)

(What is life? A frenzy.
What is life? An illusion,
a shadow, a fiction,
and the greatest good is small;
for all life is a dream,
and dreams are a dream.) (*Life is a Dream*)

Ideology apart, then, the very tune-patterns of such speeches were effective enough, while new literary genius delayed its arrival, to hold an audience spellbound on the level of euphony alone. They became a means of relating to traditional splendour and therefore the means of furthering mental security. The general public, in times of literary decadence, responds to outer brightness of sound and scenic movement. Until the *vulgo* could accustom itself to the starkness of non-poetical or non-emotional reasoning, not to say the chilling suspension of judgment, it needed the reassurance which inferior imitators of Golden-Age methods could provide in the energy of some of their sentiments and in most of their sound-patterns when magnified by the human voice and, increasingly, enlivened by the accompaniment of scenic display. Sound, we must remember is a force of emotional and therefore social influence. Few members of any Elizabethan or Spanish Golden-Age audience would understand every word of an Elizabethan or Golden-Age drama. Few would follow more than the general lines of Elizabethan or Golden-Age philosophy and poetic argument on the stage. Audiences were guided by sound-directives to emotionally social interpretations.

There is almost nothing in new dramas of early decades of the eighteenth century liable to alert *vulgo*-audiences to a recognition, either welcome or unwelcome, of the changes in mental values which, within limited circles of Spain, were not unknown. The popular dramatists and, by extension, their audiences, represent popular assumptions, many of which could and would be defended by opponents of the *Ilustristas* and many of which would eventually have to be discarded, however unwittingly.

Two of the most popular dramatists of the early decades, Antonio Zamora and José de Cañizares, were also the most artistic in their conservatism. Philosophical, social, and sentimental attitudes were transferred to their plays from tradition, sometimes in conscious emulation of their models, sometimes instinctively. And this, despite the fact that the intelligent Zamora, who had become, as stated in an edition of his plays in 1722, a 'gentilhombre de la Casa del Rey' (Gentleman of the King's Household) and 'Oficial de la Secretaría de las Indias' (Officer of the Secretariat of the Indies), must have been within mental access of untraditional, Bourbon sources or change of restlessness. The 1722 edition of his plays is dedicated to the conspicuously unSpanish minister Grimaldi. But Zamora, at the time of the Bourbon accession, was a man of some forty years, maturely set in national techniques of dramatic sentiment and presentation which he deliberately 'imitated', he said, from Calderón, the 'mayor maestro de este arte difícil' (greatest master of this difficult art).[1] Therefore, none of his right-minded characters is inspired to doubt the validity of pre-Newtonian codes of values, to suspend his judgment over ambiguities, or to resort to rationalist forms of expression. Nor, significantly, does anyone experience the need positively to defend conservationist values. In the Zamoran world, the public world of the theatre and of popular expression of popular belief, no danger to settled social assumptions had yet intruded. His heroes follow in the sure steps of Calderonian heroes and are distinguished from his only in that their philosophy is less cogently expressed, that their self-assured rhetoric tends to deteriorate into rant, that their sense of humour is less evident, and that they and all other characters depend more on sensational occurrences and the concrete support of stage-machinery. Zamora was not technically inartistic, but his ideology derived from thought-solutions made by past masters of past thinking in a past time when Spain was sure of her authority. Old forms for a while can continue to stand in for thought. Old forms for a while can represent to the *vulgo* a theoretical, if not practical standard of life.

Zamora, therefore, eloquently promoted sounds and outward

signs of emotional continuity which left his *vulgo* undisturbed, the more necessarily, naturally, and appreciatively so, perhaps, in circumstances of dynastic change. His phraseology, attuned to inherited sentiment, was stirring enough to 'inspire' the average orator and popular writer both with stimulating memories of Spanish grandeur and, in the absence of an inner complexity of ideas, with the skeletal means of reproducing theatrical sound-suggestion. That triumphant Calderonian sound-climax, drawing together key-words of parallel arguments that have been distributed over various lines of parallel form and are now reunited as a finality of solution in the last sentence, is regularly copied by Zamora, even incidentally:

Lizana: ¿Tiene el amor
para que las almas fleche
mejor harpón? ¿Tiene el mayo
para adornar sus vergeles
mejor flor? ¿Mejor lucero
tiene el espacio celeste
de la esfera? No. ¿Pues cómo
pude amar otra, si excede
en hermosura a deidades
estrellas y rosicleres?
(*Cada uno es linaje aparte*, Act I)[2]

(*Lizana*: Has love a better harpoon for spearing souls? Has May a better flower to adorn its orchard fields? Has the celestial stretch of the sphere a better shining star? No. Then how could I love anyone else if she exceeds in beauty the deities, stars and the shining of dawn?)
(*Everybody Is of His Own Lineage*)

The verbal paradoxes, which once had suggested paradoxical thought but now, by indiscriminate overplay, suggest merely two opposite ways of saying the same proverbial thing, owe their conception to a long playhouse-tradition of teasing the audience into thoughtfulness, or at least into the expectation of alluring mysteries to come. Zamora's tone of heroic patriotism, neither aggressive, nor defensive, is complacently self-assured. His series of rhetorical questions are derived from his predecessors' wealth of emotional possibilities, and now largely repetitive:

Capistrano: ¿Qué me quieres, fantasía?
Discurso, di, ¿qué me quieres?
¿No le basta a mi fortuna
la mudanza de mi suerte,
sino que el entendimiento
en batallas más crueles
padezca también la lucha
de imaginadas especies,
ya en iras, que me enfurezcan
ya en celos, que . . .? Tente, tente

memoria . . . .
¿Cómo es posible que el noble
generoso ánimo fuerte
de mi corazón, consienta
que la vilbárbara plebe
mis esfuerzos aprisione,
y mi osadía sujete?
¿Cómo es posible . . .?
¿Cómo es posible?
¿Cómo podrá, Cielo Santo . . .? etc.
(*San Juan Capistrano*, Act I)[3]

(*Capistrano*: Fantasy, what do you want of me? Tell me, discourse, what do you want of me? Is it not enough for my fortune that my lot should change, without having my understanding, in most cruel battle, suffering the strife entailed in imagined species, now in forms of anger that madden me, now in jealousy, which . . .? Hold, hold, my memory . . . How is it possible that the noble, the generous and strong spirit of my heart should permit that my forces should be imprisoned by the basely barbarous plebs and my audacity subdued? How is it possible . . .? How is it possible? Holy Heaven, how could . . .?)

For the actor are provided shouting techniques to catch and hold attention by the repeated use of the same parts of speech, notably imperative verbs, or a succession of domineering nouns and pronouns, which allow his voice to rise by convenient stages on successive rungs of the voice-ladder:

Belerofonte: Vandido monstruo . . .
Sal . . . .
Sal . . . .
tu ocio cobarde, tu traidor sosiego . . . . .
embiste, asalta, lidia, pues te espera
amante guerreador . . . . .
¡O! iqué en vano tu cólera, Quimera! . . . .
(*Todo lo vence el amor*, Act III)[4]

(*Belerofonte*: You monster bandit . . .
Out . . . .
Out . . . .
your cowardly leisure, your treacherous ease . . .
attack, assault, fight, for there awaits you
a warrior lover . . .
Oh how in vain is your anger, Chimera!)
(*Love Conquers All*)

or

Laurencia: Esto es
la furia, el dolor, la rabia,
el despecho, el frenesí
de una mujer desdichada.
(*San Juan Capistrano*, Act III)[5]

(*Laurencia*: This is
the fury, the sorrow, the anger,
the despair, the frenzy
of an unhappy woman.)

The grammatical structure of descriptive passages are soothing in a *continuo* of companiableness, like background and introductory music directing the atmosphere of modern films, or the ritual music sung or played in Church services of which the sermon is a ritual part:

Marte: Sufra, padezca
Júpiter: Gima, y lamente
Marte: Tus iras crueles
Júpiter: Tus ceños violentos
Los dos: Y toquen a marcha esferas y vientos.
Arma, arma, guerra, guerra.
Guerra, guerra, arma, arma.
. . . .
Músicos: Arma, arma, guerra, guerra,
guerra, guerra, arma, arma . . .
(*Todo lo vence el amor*, Act I)[6]

(*Mars*: Let [that one] suffer, sorrow
Jupiter: Moan and wail
Mars: Your cruel wrath
Jupiter: Your violent frowns
Both: And let all spheres and winds be called to action.
Arms, arms, war, war.
War, war, arms, arms . . .
Musicians: Arms, arms, war, war,
War, war, arms, arms . . .)
(*Love Conquers All*)

All these are sound-customs which had a practical social value, and popular writers, orators, and actors of the early eighteenth century would not easily be divorced from their well tried, indeed highly effective means of mass-communication. The actor who can increase the spirit of tension by subtleties: by a conspicuous stillness of physical reserve; by unobtrusive movement, delicately nervous changes of tone, facial expressions indicative of conscious self-control, was not encouraged to develop in histrionic genius by the power of suggestion until the twentieth century. An eighteenth-century public looked for concrete explicitness, and would have been as uneasy in the company of those uncertain powers of implication as it was in the company of *Epoche*.

If Zamora occasionally satirizes social customs in the interests of comic relief, he does so only within Calderonian limits. For instance, his avuncular jibes at the ignorance of pretentious students, among them medical students, as in *El hechizado por fuerza* (*The Man*

Bewitched by Force), where they disagree over treatments and resort to the safety of ill-understood medical jargon, are not criticisms of medical practice in Martínez's sense of scepticism, but are the ever popular borrowings from Golden-Age dramatic practice and are meant to provoke homely amusement at family failings rather than the double-edged laughter of prospective reformers. When interpreting words one must relate them to the tone of their context. So, too, Zamora's serious characters, his heroes, heroines and villains, historic or contemporary, his 'crowds' and minor persons, conform to standards long established in the *corrales* and automatically accepted by theatre-goers. He may be seen in his times as a model for public attitudinizing.

Zamora's younger contemporary and compeer, José de Cañizares, who died in 1750, seems better circumstanced by his age to appreciate the possibilities of change. Certainly he was a less craven imitator of Golden-Age heroics than Zamora. Yet for the most artistic of his serious scenes he relied no less than did Zamora on traditional stage-mannerism and oratory. If, in response to Gallic taunts of Spanish stagnation, he turned for certain themes to Greek and Roman classics, it was to castilianize their forms of expression.[7] There is no denying that one of Cañizares' best examples of craftmanship, *El dómine Lucas* (*The Master Lucas*),[8] is an intelligent caricature of windy heroics and rhetorical pedantry; that it reflects enlightened contemporary criticism of stage and pulpit-exaggerations. Truth, however, is not invariably straightforward in its manifestations. Its appearances can be partially misleading, and, in this instance, caricature in *El dómine Lucas* is less significant from the standpoint of the Enlightened than might at first appear. For, in his relatively quieter way, Cañizares, as in *Carlos V sobre Túnez* (*Charles V at Tunis*), was as faithful a recorder of rhetorical harmonies as any other pupil of Calderón. Like his Golden-Age models he had his permissively *gracioso* (clowning) moods in which, for the fun of the moment, nothing save religion, was sacrosanct, and in which he, like them, mimicked all set forms of speech. The difference between mental attitudes of *El dómine Lucas* and of Isla's *Fray Gerundio* is a difference of tone, and the force of intention. *El dómine Lucas* is set in the Calderonian mood of family fun, and the effect of Cañizares' intention is casual and lacking in malice. Neither this play, nor any other of Cañizares in which the author mocks professional standards, as in *El asturiano de Madrid* (*The Madrid Asturian*):

> *Blas*: [referring to a medical doctor]¿Luego mata aquel señor?
> Es verdad, más con licencia. (Act II)[9]

> (*Blas*: Then did that gentleman kill?
> In fact he did, but with the licence [to do so].)

would detract from an audience's comforting reliance on emotional declamation, or family teasing, superficially reminiscent of Calderón. If there is a serious intention or message behind *El dómine Lucas* it is not concerned with a desire to banish rhetoric but to improve it. Cañizares is nearly as Calderonian as Torres Villarroel is Quevedan. Both escape from the restriction of fully committed, empiricist reasoning by the compromise of paradox in which large appearance can be made to carry a semblance of a large thought. The average orator of the day might not find in *El dómine Lucas* an excuse for heroic declamation, but he could safely startle his audience into attention by convenient parallels to Calderón's dramatic paradoxes, such as:

> Vos a estudiar en la guerra
> yo a lidiar en los estudios
> (*El dómine Lucas*, Act I)
>
> (You to study in war
> I to fight in studies.)

which implies depth but does not involve a change or loss of ideological direction.

In spirit, at least, the average, serious orator who responded to and catered for the general need of dramatic colour and readily understandable, sensory illustration, conveyed through allegory and mythology, was necessarily competing with authors and actors not only of heroic dramas but of *autos sacramentales* and *comedias de magia* (mystery plays and plays of magic), which formed part of the *vulgo*-media of expression. And it will be remembered that the word *vulgo* in eighteenth-century, intellectual parlance, need have no social implication, and must be equated with the term 'general public'. One of the most notoriously popular plays of the period was the five-part *El mágico de Salerno* (*The Magician of Salerno*), of 1733, written by a contemporary of Cañizares, Salvo y Vela, and republished and replayed until the end of the century. Its importance for our present purposes consists in its author's understanding of his public's love of illustrative wonders. The average orator wishing, for moral or other reasons, to arrest public attention, tended to follow in the psychological wake of stage-successes, and brighten his argument with illustrations of miracles or marvels paralleling those performed in the theatres. Simplification by allegory and parable is an obvious method of teaching. When such illustration is brightly painted its effect is increased. In *El mágico de Salerno* rightminded characters triumph by sensational means. The heroine Diana is saved at the stake, through white magic, by the transformation of the stake into a flower stand. The mixture of Classical characters, such as Ceres, with allegorical characters, such as La Abundancia (Plenty), derives from a mixture of literary and folk-traditions well publicized on the stage and contagiously affecting those who knew how to speak to the *vulgo* on a *vulgo*-level. This wonder-play, with its wealth of transformation-scenes and storied argument, was one of the many which encouraged *vulgo*-taste for clamorous complexity. Not that orators were invariably aware of all their sources of inspiration. Like their public they were usually conditioned by inherited circumstance.

Nowhere is the social classlessness of *vulgo*-taste better exemplified than in Fernando de Barcena y Orango's *La Babilonia de Europa y primer rey de romanos* (*The Babylon of Europe and the First King of the Romans*), 'escrita para el . . . Duque de Osuna' (written for the. . .Duke of Osuna), in an edition of 1731.[10] At a time when Cañizares might be said to range highest in popular dramatic artistry, his fervent admirer, Barcena Orango, must be placed lowest. *La Babilonia de Europa* dedicated to his patron, the Duke, with incidental praise of the 'great' Cañizares, and professions – truer than the writer intended – of his own artistic unworthiness, was apparently meant initially for performance in the Duke's household: an aristocratic circumstance not uncommon in eighteenth-century Europe. Private productions of specially commissioned, or selected dramas, often performed by members of an aristocratic household, could often afford a playwright greater opportunities for enriching scenic display, possibly out of doors, than did productions in the *corrales*. At all events, this playwright, assuming that neither aristocratic nor commoner attention harboured discriminating qualities of rationalism, that audiences in general must be captured and held spellbound by spectacular action and not required calculatingly to reason, concentrated his energies on physical excitement. There, on the lowest plains of popular drama, his characters, including Rómulo, King of Rome, the gallants, Sabine *damas* (ladies), the gods Neptune and Jupiter (the last mounted on an eagle), the surging crowds and soldiers, are involved, with *pundonor*-(point of honour-)incidentals of an incidental *comedia de enredo* (drama of intrigue), in storms, battles, warcries and war-trumpeting, Roman festival activities of bullfighting, and song. On the popular heights, where Zamora and Cañizares trod, these artistically cheaper – because most easily obtainable – means of manufacturing tension and holding spectators in thrall, were used much more judiciously. But the need of the classless *vulgo*, in any place or time, is for tensions manipulated externally, for perceptible stimuli comparable with the need for miracles and marvels in religious, or superstitious, practice, or for dramatized oratory in preaching. The demand for practical excitement in customary recreations or for any form of public expression and display has always been a force with which reformers have had to reckon. Faced at times with this natural force the *ilustrados* recognized one of their strongest adversaries.

From theatrical traditions of sound and visual patterns, faithfully preserved by a Zamora and Cañizares, or, on a lower level, by Salvo y Vela and Barcena y Orangos, the average orator in Church or elsewhere, in person or in print, instinctively took his bearings. When he exaggerated their heroic pitch and volume for his own immediate purposes, he had still not descended to his lowest artistic or religious level. Religious sound and fury to the general public can at best be stimulating. The orator was at his worst, by enlightened standards, when he fell, understandably, but unfortunately, under the influence of playwrights who enlivened religious themes, especially, with the *vulgo*-throb of sentimentality: a sentimentality not as yet inspired by European interest in the human pathos of commoners, but a sentimentality accidentally produced by inadequacy, by the inability of lesser dramatists to present any kind of homely emotion without melodrama or bathos. The understandable need was for new models of homely illustration, from homely sources, capable of suggesting homely parallels and bringing worthier tears to the eyes of susceptible congregations.

In such circumstances one of the most interesting, if most inartistic, plays of the early decades is the two-part *La coronista [sic] más grande de la más grande historia. Sor Ma. Jesús de Agreda (The Greatest Chronicler of the Greatest History, Sister Mary Jesus of Agreda)*, of 1736, written by Manuel Francisco de Armesto, who describes himself as 'Secretario del Secreto de la Santa Inquisición' (Secretary of the Secret [Service] of the Holy Inquisition). This, therefore, was a play which a representative of an accredited Establishment was offering as suitable dramatic material for religious purposes, though, in fact, its travesties of human feeling, emphasized with the best of human intentions, were precisely those which, transferred to sermons, afforded Father Isla the greatest disgust.

La coronista más grande . . ., in its would-be picturesqueness, encloses a series of sermonets delivered at strategic intervals by Sor María on the life of the Virgin, and set into a sensational love-story with as many trimmings of illustrative magic and marvel as any drama could physically accommodate. The bid for spectators' attention and emotional cooperation determines every dramatic mood and verbal sound. One of the easiest, most hackneyed, and yet most effective ways of appealing to *vulgo*-emotion, as sensational writers of the *drame* were later to discover,[11] was to introduce children into the adult action, and over-emphasize either the moving wisdom of their simplicity, or the moving vulnerability of their innocence. Here, in visions periodically experienced by Sor María de Jesús, and materialized on-stage for the audience's benefit, her namesake the Virgin Mary appears as a child, incongruously calling Sor María 'hija' (daughter), and Sor María, in her turn, addressing the child Mary as 'Madre' (Mother).[12] Elsewhere, in a Christmas scene, both Jesus and His mother appear as infants together, though this is not a play meant for child-audiences.[13] On the contrary, such actions occur in a mixture of heroic and *capa-y-espada* (cloak-and-sword) settings. Also somewhat mawkishly, not to say somewhat irrationally, when the child Jesus shows to Sor María the signs of His Passion and talks about His crucifixion, they address each other as 'Esposo' and 'Esposa' (Husband and Wife).[14] Obviously it was intended that when this play was performed there should not be a dry eye in the theatre, and that the 'sermons' on the life of the Virgin, delivered, under inspiration, by her chronicler Sor María, should be sensually impressed. No wonder was it that other forms of preaching should have borrowed similarly effective imaginings and illustrations; that the popular preachers' tones should have reproduced the sob-sounds, the *innuendo*, the thunder of drama – Armesto's play, in fact, begins with an earthquake; that such preachers should have appealed to imaginative congregations with examples of miracle and marvel; that they should have aped *gracioso*-relief by an occasional, humorous aside, however devotionally irrelevant. No wonder was it, therefore, in times when lively, heroic activity and stage-surprises were as generally attractive to public taste as is modern western- or science-fiction, that the orator tended to enforce his point and convey his message vividly by means of theatrical story-telling. Sor María, with her visions and levitations – presumably manoeuvred by up-to-date machinery – moves within the framework of a *pundonor* love-story external to herself, but eventually influenced by her for good. In other words, her magic and mystery are dramatically evolved within a stagily human community of *caballeros*, *damas* (knights, ladies) and clowns. An accompaniment of music for the most serious scenes, such music as was encouraged even as early as the first decade of the eighteenth century by the vogue for Italian opera, may not have been available to the preacher in the pulpit, but it was available as a suggestive adjunct in street-performances of the *autos* (religious plays), and would suggest to orators and preachers the importance of melodious relief-tones after thunderous declamation, and of occasional quotation from lyrical poets.

A curious feature of Armesto's published edition is the *Aprobación*, or formal sanction, given to the play in 1735 by Antonio Téllez de Azevedo. This editorial judge was evidently alive to French criticism of Spanish drama, for he refers to the importance of the

'Rules of Art' and deplores *vulgo*-taste for sensationalism. What is surprising, therefore, is his wholesale approval of a drama which is not merely lacking in the Unities, but which contains as much sensationalism of natural and unnatural phenomena, dancing, fighting, and scenic spectacle in one act as most popular dramas presented in three.

Don Antonio's standard of values, of course, was confused. This artist, he writes, has known how to:

> . . . balancear de suerte los extremos, que ceñido a las precisas Reglas, ha conseguido, con la principal aceptación de los Doctos, enlazar la común diversión de los ignorantes, mereciéndose de todos el mayor aprecio.[15]
>
> (. . . so balance the extremes, that, while observing the exact Rules, he has managed with the notable approval of the Learned to relate to the general amusement of the ignorant, earning from all the greatest appreciation.)

Which means that Don Antonio, who thought it internationally intelligent to talk of the Rules and decry the *vulgo*, was instinctively bending those Rules to established Spanish standards, his own standards, be it noted, not less than those of his fellow members of the national *vulgo*. His attitude has significance in that it represents a fundamentally uncritical acceptance of what was familiar: a general attitude as natural then as, in equivalent circumstances, it would be now. This thoroughly inartistic play is nevertheless vivid in colour and stimulating in its popular emotion. The episodes and fulsome sentimentality would certainly infect a *vulgo*-audience, including *vulgo*-clergy, with theatrical interpretations of reality. It must be remembered that on the stage there was little distinction anywhere as yet between artistic appeal to the masses and artistic appeal to the discriminating few. The only difference between the technique of appeal in Golden-Age drama to the masses and that in early eighteenth-century drama to the similar masses was that Golden-Age technique was constructed with innate genius and that of the early eighteenth century was not. For the instruction of the masses it seemed natural to preserve and so, in the absence of genius, to exaggerate familiar forms of public address, and to oppose innovations advocated by those who put more emphasis on plain reason than on imaginative feeling. Often too, in practice, as Isla himself exemplifies, reason confusingly defeated itself by stagey rhetoric with which some of its own precepts were advertised.

Maintaining the close, personal and practical relationship of tradition between Church and theatre, the Chaplain to the Real Monasterio de la Encarnación, Tomás de Añorbe y Corregel,[16] produced in the 'thirties a string of popular dramas, sacred and profane, that might serve as models for the art of riveting *vulgo*-

attention. Añorbe was not a name deserving of recognition in literary annals, but he was not devoid of imagination and knew how to emphasize, indeed over-emphasize, the more obviously sensational techniques of the Golden Age. He had vehemently defended himself, in conjunction with Golden-Age writers, especially Lope de Vega, the 'mejor cisne castellano' (the finest swan of Castile),[17] in a Preface to his *zarzuela*, *Júpiter y Danae* (operetta, *Jupiter and Danae*), in 1738, against a criticism in *El Diario de los literatos* of 1737.[18] With an aptness, more ironical than he realized, he had quoted the lines of Lope's goodnatured argument that strict rules of art might be all very well in their own way, and in certain circumstances, but that his own plays:

> aunque fueran mejor de otra manera
> no tuvieran el gusto que han tenido.[19]
>
> (although they might be better [written] in another way
> they would not make such a good effect.)

The same was true of all forms of address to the general public: a fact with which every Martínez, Feijoo and Isla would have to contend. Añorbe himself, ironically and unwittingly, proved Lope's case when, either in a repentant moment, or, more likely, in a piqued desire to show that he could observe the Rules of Art with the best, he produced his pitifully enervating tragedy *El Paulino* (*Paulinus*) of 1740, based, he says, on Corneille's *Cinna*. Even here, however, where he is reduced to constrained reasoning, and cannot call to his aid the dramatic action of florid marvels of surprise which quicken an audience's pulse-rate, he resorts to the tense question-and-answer rhetoric which sober orators and preachers, such as Isla or Massillon, rely upon to recall lapsing attention, making use of varieties of tone and pitch from shout to whisper. Rant of itself can be publicly effective. Yet rant has more public appeal when it is related to dramatic story-telling, with, for instance, improbable parallels of human paradox to sting imagination into activity, and with a warmth of lavish illustration.

A rich source-book of technique for popular sermons and oratory is Añorbe's three-part *La tutora de la iglesia y doctora de la ley* (*The Woman Tutor of the Church and Woman Doctor of the Law*), of 1737, built round biblical incidents in the life of Jesus and sentimentalizing the Bible story, sometimes to the point of the ridiculous, but not less so than the popular sermons already quoted. Such puerile examples as King Abagaro's sympathetic message to the vilified Jesus: 'Aquí tengo una ciudad, pequeña es, y honesta, pero al fin para los dos bastará' (I have here a city, which though small and modest, yet altogether for the two of us will suffice),[20] could, when delivered in the right tone, move a general audience or a mixed congregation to tenderness. Many printed sermons rely on touching illustrations of the same kind. Moreover Añorbe, like Armesto, but often with more imaginative artistry, employs the stage-force of vocative address. This technique is commonly used to vitalize oratory especially when accompanied by creative adjectives which, at their best, give personality to an inanimate object:

Negra, macilenta noche[21]

(Black, wan night)

for example, is a line worthy of the oratorical Torres Villarroel. And other Baroque orators who appreciated the occasional verbal paradox of Gongoristic memory, and the value of mixing the sublime and the grotesque for capturing the public's straying attention, had a convenient reference-book in the shape of Añorbe's *Los amantes de Salerno* (*The Lovers of Salerno*). His dramatic vocatives at moments of suspense:

Lisandro: Aves de la noche triste,
melancólica ciprés,
opaca luz macilenta,
palacio, injusto Babel,
y tú, jazmín deshojado,
triste cárdeno clavel
sed testigos . . . (Act III)

(*Lisandro*: Birds of the sad night,
melancholy cypress,
dim, wan light,
unfair Babel of a palace,
and thou, leafless jasmine,
sad, cardinal red carnation
be witnesses . . .)

however unworthy of Calderonian tradition, maintain the pattern of changing voice-pitches and bequeath to the world of oratory, if not memorable sentiments or images, at least the memory of an inspiring variety of rant-tones and the possibilities afforded to the voice by the repetition of dominating grammatical structures.

The influential strength of stage-oratory, decadent as was its eighteenth-century expression, was maintained for the general public in later decades by dramatists as ideologically uninspired as A. Bazo, L. Moncín, L. F. Laviano and their like,[22] and was eventually superseded by a more creatively literary, verbal appeal to public emotion in the exhibitionism of Metastasian opera and a new brand of international heroics. Occasionally the florid successors of Añorbes, Salvo y Velas and Barcena y Orangos reflect fleetingly some slight awareness of a new talk of reasoning. Antonio Bazo, for instance, a notable purveyor of empty sound and fury in such typical works as *Los tres mayores portentos en tres distintas edades. El orígen religioso, y blasón carmelitano* (*The Three Greatest Portents in Three Different Ages. The Religious Origin, and Carmelite Heraldry*), stopped momentarily for some semblance of thought in his *Sastre, rey, y reo a un tiempo, el sastre de Astracán* (*Tailor King and Culprit [King] at One and the Same Time, the Tailor of Astrakhan*), set in Oriental scenery, and written under the pseudonym Antonio Furmento.[23] In Act I of the absurdly impossible plot of Bazo's possibly worst drama, the 'sastre', supposed son of a 'sastre', is chided by his 'father' for an outburst of feeling:

Xouran: ¿. . . que de refrenar no tratas
tan altiva condición
arrogancia tan extraña?

(*Xouran*: . . . so you don't try to restrain
such a haughty disposition
such outlandish arrogance?)

and the 'son' objects to homely 'reasoning' as a substitute for the superiority of unobvious instinct:

Schenedin: No des tal nombre, Señor,
al aliento que me inflama
. . . .
pretendo darte a entender,
que no poco me avasalla
la razón, cuando hasta aquí
guarde, sin que al labio salga,
en el alma la altivez,
en el pecho la arrogancia . . .
. . . . quisiste
que dejara aquella instancia
para que en tu oficio bajo . . .
enfrenaran mi soberbia
tareas tan ordinarias
haciendo que a vil aguja
trocase la noble lanza . . . (Act I)

(*Schenedin*: Don't give such a name, Sir,
to the breath which inflames me . . .
I want you to understand
that I am not a little enslaved
by reason when up till now
I hold back, without letting it reach my lips
the presumptuousness in my soul
the arrogance in my breast . . .
. . . . you wanted
me to leave that matter
in order that in your low office . . .

my pride should be restrained
by my changing for a vile needle
the noble lance . . .)

An audience would have been in heartfelt agreement with him. But perhaps the very fact that two disputants might seem disposed to argue psychologically is significant of how drama might be expected one day to take a different turn. Incidentally, it is noteworthy that, as yet, there is no hint in Bazo's technique of the emotional dramatization of everyday circumstances of commoner life which towards the end of the century abounded in sentimental plays. This 'sastre'-king despises his *sastre*-background that he might respond with colour and excitement to the *vulgo*'s demand. As yet political atmosphere abroad and at home had not been affected sufficiently by intimations of social revolution.

It takes time to train any general public to appreciate strict verisimilitude either in stage-circumstance or dialogue. When reviewing Alarcón's *La crueldad por el honor* (*Cruelty for Honour*), the *Diario de los literatos* in all honesty could not withhold praise for the author's theatrical ingenuity. But praise is tinged with blame for audiences' expectations of excitement, expectations typical, one might add, of *vulgo*-audiences in any century including the twentieth:

> . . . la disposición de los lances es muy ingeniosa, y acomodada al gusto de la Nación, que le divierte más lo admirable que lo verisímil.[24]

> (. . . the arrangement of the incidents is very ingenious, and is accommodated to the taste of the nation, which is entertained more by what astonishes than by what is true to life.)

In general, then, and below the surface of any new tendency towards cold or constrained intellectual rationalism and plain speaking, the popular warmth of *vulgo*-oratory ingloriously survived, and eventually was brought into a psychologically higher life by the literary Romanticism of the nineteenth century. From the standpoint of the general populace, preachers and all public orators could best impress by impressive declamation, and the difficulties of the Enlightened were accordingly intensified by the fact that they spoke to home-audiences in an alien language unaccompanied by histrionics.[25]

NOTES

1 Antonio de Zamora, *Comedias nuevas con los mismos sainetes con que se ejecutaron*, (Madrid, 1722), 2 vols. See Vol. I, 'Prólogo', n.p. For a discussion of Zamora's drama in general see Paul Merimée, *L'Art Dramatique en Espagne dans la Première Moitié du XVIIIe Siècle*, 2nd edition (Toulouse: France-Ibérie Recherche, 1983), pp. 23-80.

2 *Comedias*, Vol. II (Madrid: Joaquín Sánchez, 1744).

3 *Comedias nuevas*, *ed.cit.*, Vol. I. The full title is *El custodio de la [H]ungría, San Juan Capistrano.*

4 *Comedias nuevas*, *ed.cit.*, Vol. I.

5 *Ed.cit.*, Vol. I.

6 *Ed.cit.*, Vol. I.

7 See McClelland, *Spanish Drama of Pathos*, Vol. I, p. 116.

8 I have used the undated edition in the Biblioteca Nacional. See also BAE, XLIX (*Dramáticos posteriores a Lope de Vega*, II).

9 *Comedias sueltas*, 3 vols. Vol. III, n.d. G. n. 8.

10 *Descripción cómica. La Babilonia de Europa y primer rey de romanos* (Madrid: Gerónimo Roxo, 1731). B.Nac. T.1.129.

11 See McClelland, *Spanish Drama of Pathos*, Vol. II, p. 478 *et passim.*

12 See Part I, Act I.

13 *Op.cit.*, Part I, Act III.

14 *Op.cit.*, Part I, Act I. Concerning *vulgo*-extension of the Marian cult, see Joël Saugnieux, 'Ilustración católica y religiosidad popular: el culto mariano en la España del siglo XVIII', in *La época de Fernando VI* (Oviedo: Cátedra Feijoo, 1981), pp. 275 ff. In general relation to the conception of religious expression as a public 'diversion', see also Julián Martín Abad, *Contribución a la bibliografía salmantina del siglo XVIII: la oración sagrada* (Salamanca: Ediciones Universidad de Salamanca, 1982), p. 36 *et passim.*

15 See the edition of Madrid, 1736. The Aprobación, dated 1735, has no pagination.

16 See the title page of *Júpiter y Danae* (Madrid, 1738).

17 See his 'Prólogo apologético' to *Júpiter y Danae*, *ed.cit.* There is no pagination.

18 *Diario de los literatos*, Vol. IV, 1737, *Artículo* XVII, on Añorbe's *Comedia de la tutora de la iglesia.* See I. L. McClelland, *The Origins of the Romantic Movement in Spain*, 2nd edition (Liverpool: Liverpool University Press, 1975), pp. 12ff.

19 See Lope's *Arte nuevo de hacer comedias*, ed. H. J. Chaytor, *Dramatic Theory in Spain* (Cambridge: Cambridge University Press, 1925), p. 15.

20 See Part I, Act I.

21 See *Los amantes de Salerno*, n.d., Act I.

22 See McClelland, *Spanish Drama of Pathos*, Vol. II, pp. 546 ff.

23 See F. Aguilar Piñal, *Bibliografía de autores españoles del siglo XVIII* (Madrid: C.S.I.C., 1981), Vol. I, p. 555.

24 *Diario de los literatos . . .*, Vol. I, 1737, *Artículo* IV, p. 81.

25 On the subject of the popular importance of vocal theatricality see also: Ermanno Caldera, 'Entre cuadro y tramoya', *Dieciocho*, IX (1986), Nos. 1-2, 51-61.

CHAPTER 6

Disturbing Effects of the Periodical Press

For at least the first two decades of the eighteenth century the Spanish Periodical Press did not act as a direct disturber of established ideas, and, in the next decade, the clandestine *El Duende de Madrid* (*The Madrid Elf*), dealing with political matters, circulated too privately to be more than privately influential.[1] Branches of free-range journalism directed to the general public had not yet taken organized form. The conception of the Periodical Press was that of state-controlled, reporting newspapers, or, for instance, of astrological almanacs, also regarded by the general reader as unquestionable reports of fact. So far public journalism had not been diverted into grey, side-areas of debatable supposition, and, therefore, unease; and the Periodical Press in general had not yet developed ideological character, least of all critical individuality. Analytical journalism, both subjective and objective, and which the general public was not yet trained to assimilate, was to emerge, provocatively, from outside regular sources of state pronouncements.

Strict state-censorship of the early decades, from 1701, confining the weekly *Gaceta de Madrid*, and, by extension, its provincial counterparts, to approved news-items, was a natural war-time and post-war precaution. In these newspapers the statements of politically national news are paternally concerned with Spanish successes in war, or with the health, movements and activities in general of the Royal Family. They refer to incidents such as the King's warm welcome to the Jacobite Pretender, the signing of the Quadruple Alliance Treaty, the decision to remove Alberoni from power in the interests of peace. Or they make family references, for instance to the King's method of teaching Prince Luis to gain experience by allowing him to work in the Royal Office, or to the slight cold, 'un poco de calentura', suffered by the Prince of Wales in London.[2] Political home affairs, as reported so guardedly, were unlikely, then, to promote the growth of intellectual disturbance. Nevertheless, there were certain aspects of the *Gaceta*'s impersonal reporting which, even unwittingly, could act as sources of changing ideas, and so promote unease. The *Gaceta* announced new publications, not merely the variety of over-popular almanacs, but, amidst the increasingly crowding publications of the 'twenties, the *Medicina scéptica* of Martínez[3] and, later, the work of such outstandingly self-assertive writers as Feijoo, Torres Villarroel, and their critics or supporters. In other words, the *Gaceta*'s announcements promoted library-suggestions. Still more significant, in effect, would be the impassive report, in intent, of foreign affairs, often given extensively; and of topical phenomena at home and abroad. Especially suggestive, in view of those announcements of new and controversial surveys of Medicine, Science, and other subjects relevant to practical living, are the sometimes obsessively repeated reports on, for instance, the French plague (15.X.20); treatment of the Pope's illness (8.IV.21); and the King's smallpox (29.VIII.24, 5.IX.24); abnormal phenomena such as the birth of quadruplets to a woman in Madrid (15.X.20), or the Mallorca floods (29.IV.21) and other weather peculiarities.[4] Not all the titles of the books and articles announced suggested academic exclusiveness. Attention would be drawn to works of obvious popular appeal, like the *Modo de aplicar los remedios a todo género de enfermedades en ausencia de los médicos* (*Method of applying remedies to every kind of illness in the absence of doctors*), by Hermanos de la Congregación del Padre Bernardino de Obregón (Brothers of the Order of the Father Bernardino de Obregon) or *Floresta astrológica, perpetua y verdadera que contiene una breve y segura explicación de todos los elementos* (*Astrological grove ever-lasting and true, containing a short and certain explanation of all the elements*).[5] Readers of average curiosity would eventually be likely to encounter, even in popular folk-evaluations of Medicine or Science, those references to rationalistic doubt that fostered academic wars. News from abroad would be likely to invite Spanish/foreign comparisons. In a minor way, then, the *Gaceta* was a supplier of potential means of recognizing ideological change and testing its implications. From the 'thirties onwards the number of publications, Spanish and foreign, announced by the *Gaceta* notably increased. Not that the *Gaceta* itself engaged in reviews of books thus announced. Nor, significantly, was it yet assumed in most new forms of journalism, outside newspapers, and with the notable exception of the *Diario de los literatos* of 1737, that book reviews should necessarily involve evaluative criticism. So the enterprising *Mercurio literario, o Memorias sobre todo Género de Ciencias y Artes . . .* (*Literary Mercury, or Notes on Every Branch of Science and Arts . . .*) of 1739, when reporting, now with an international air of calculated broadmindedness, on criticisms both used by and evoked by the *Diario de los literatos*, sees fit to mention the principles of 'nuestro instituto, en que nos obligamos a no usar de expresión alguna que tenga la menor apariencia de crítica' (our institution in which we undertake to use no expression which may have the slightest appearance of criticism).[6]

The fact was that the *Mercurio literario . . .*, which could worthily consider itself *ilustrado*, and which had praised the *Diario*'s undertaking, was in two minds about its right to brandish criticism satirically.

> . . . confiesan [los autores del *Diario*] que son satíricos, pero que tienen razón para serlo, añadiendo que en las Naciones Extranjeras donde no se escribe tan indignamente como en España, no son menos ásperas las censuras para lo que citan varios ejemplares, los que no cotejamos con los que pudiéramos citar de su *Diario*, porque no parezca que queremos manifestar su desigualdad, lo que sería faltar a nuestro instituto, en que nos obligamos a no usar de expresión alguna que tenga la menor apariencia de crítica.[7]
>
> (. . . [the authors of the *Diario*] confess that they are satirical, but that they have reason to be so, and they add that in Foreign Countries, where writing is not as poor as it is in Spain, criticism is no less harsh, of which they give various examples such as we do not compare with those which we could quote from their *Diario*, because we do not wish it to appear that we want to demonstrate their difference, for that would be failing [in our duty towards] our institution in which we undertake to use no expression which may have the slightest appearance of criticism.)

One is reminded of honest fears about moral effects of Isla's *Fray Gerundio*, or of Feijoo's satirical laughter. Not, of course, that satire was new to Spanish experience. Any reasonable disturbance now would be caused by the development of the art of satirical criticism in unexpected and unexplored directions: that is, beyond tabulated boundaries. Advanced *ilustrados* were ready to experiment, in criticism, as in science, without the traditional constraints. Bystander-fear for the future lay in the conception of possible uncontrol. Dramatically, then, in 1737, the ostentatious *ilustrismo* of the *Diario de los literatos* had acted as a literary equivalent of Martínez's scandalizing *Medicina scéptica*. Now the *Diario*-scandal consisted in the editors' concession to free-range criticism: their academic assumption that analytical criticism, negative and positive, was an essential feature of mental vitality and literary life; their elaborate formation, at an artistic level, of a new universe of critical atmosphere; and, especially, their dramatic passage from impersonal announcement of publications to a self-confident dissection of them by known individuals of individual principles. This formation, on paper, of a seeming School of General Criticism, unrestricted by traditional preconceptions, could appear to a *vulgo*-mind to favour arrogance and lead the way alarmingly to uncontrollable disruption. Practice in open-ended literary debate was in fact another means of practising openmindedness, especially in the international context of a rationalistic Press.

Winds of protest raised by the *Diario*'s assertion of its rights to uninhibited criticism cannot, of course, be attributed always to reasonable preoccupations. A blustering Pedro Nolasco de Ozejo, for instance, had taken personal offence at the *Diario*'s satirical treatment in Vol. IV of his *Vida de San Antonio Abad* (*Life of St Anthony Abbot*) written in *octavas* (eight-line stanzas).[8] And in this case the hysterically venomous tone of his reply, as he castigates the *Diario* for objecting to technical faults and clumsy neologisms – his *alternante*, *persecuente*, *requiriente* (alternating, carrying out [or pursuing], requiring [or summoning]), – or challenges its right to review works outside its professional capacity[9] and enlarges childishly and coarsely on the *Diario*'s character, would, of itself, testify to the *Diario*'s credit. It certainly takes considerable patience to thrust a way through the over self-conscious rhetoric, especially in the opening *octavas*, of Ozejo's Cantos. At the beginning of Canto I one wonders whenever the poet will come to any historical point. One tires of his obsession with would-be 'philosophical' comments on every detail of the Saint's life and with unnecessarily illustrative imagery, and high-pitched apostrophes: 'O . . . O . . . O . . .'. Exterior rhetoric, of course, is easier to copy from classical tradition than interior thinking, and therefore is all too easily exaggerated. The *Diario* had been right to regard the *octavas* disparagingly. But perhaps a more dignified, effective, and, from the standpoint of most intellectual or general readers, a more acceptable form of criticism, would have been more professionally restrained, more dismissive of the immature *octavas* than vigorously explicative. The *Diario*-critic gives the impression that in his strictures he is personally enjoying the process of demolition, and is engaged in subjective aggression.

> No se reconoce en toda la obra parte alguna, que le merezca al arte el más mínimo precepto. La erudición, sobre ser vulgar, es tan escasa que apenas se percibe; y cuando se logra es para acordar la destemplada fantasía del Poeta, que con poco acierto ofusca y confunde la idea que debían dar sólo los nombres. No parece sino que arrancó tal cual texto de la Santa Biblia para apedrear a los Lectores, confundiéndolo con el ripio que recogió en los caramanchones del Theatro de los Dioses . . .[10]

> (There is not to be found in the whole work any section which can be considered in the least way artistic. Its erudition is not only commonplace but so sketchy that it is hardly discernible: and when it emerges it is to accord with the Poet's wild fantasy, which obscures and confounds the idea which mere [noun-] names ought to give. It does not appear otherwise than that he snatched just such a text from the Holy Bible for stoning readers, confusing it with the rubble he collected in the garrets of the Theatre of the Gods . . .)

Incidentally we should notice that the writer of the *censura*, to Ozejo's reply, M. R. P. Fr. Joseph de Abadía, Lector del Convento de Santo Tomás (Assistant at the Convent of St Thomas), who speaks of present improvements – 'satisfactoria corrección' (satisfactory correction) – in Ozejo's later output, mentions the 'dureza' (harshness) of the *Diario* – critics whom Ozejo was addressing:

> ¿Quién no reparará, si cotejase este Papel de Ozejo, y otros suyos que lo acre que lleva es inducido de la dureza que halla, y le ha movido en aquéllos a quienes dirige este Papel?[11]

> (Who can fail to observe, if he compares this article of Ozejo's and others of his, that the bitterness in it is induced by the harshness he experiences, and which has influenced him in those [articles] regarding the persons to whom he addresses this article?)

Such 'dureza' in the new realm of open-ended literary criticism seemed violent, professionally and socially uninhibited and so revolutionary and disturbing. Moreover the *Diario*'s criticisms in general were not obviously categorized or expressed in the technical terms of intellectual tradition.

The Rev. Joseph de Abadía was a more reasonable thinker than Ozejo. Possibly, therefore, we should try to understand from his *censura* his conception, for his times, of critical over-emphasis.

More conspicuous among the critics and sufferers from the *Diario*'s '*dureza*' is that half-way intellectual, the Rev. Fr. Jacinto Segura, whose *Norte crítico*, first published in 1733 and republished in 1737, attempts to modernize the technique of historical research. In certain ways his semi-*ilustrismo* parallels in the world of the Arts the semi-*ilustrismo* of Torres in the World of Sciences, though any further comparison of the two would be ludicrous, for Segura had no artistic genius and little of Torres' general knowledge. Both, however, profess certain enlightened principles of scholarship which, taken out of the context of this author's interpretation, sound, for the times, rationalistically advanced. So he speaks of the importance of evidence and the historian's need to record facts without frills:

> Los que escriben con gran cuidado en la elegancia, y adorno de la oración son menos solícitos en examinar la verdad de los sucesos históricos.[12]

> (Those who write paying great attention to the elegance and adornment of stylistics are less careful about examining the truth of historical occurrences.)

He warns against quoting inaccurately from memory, against inaccurate chronological reference, misinterpretation of authors and invective criticism.[13]

Like Feijoo he could understand that there were modern, 'heretical' historians abroad, who, outside religious spheres, could be accepted as accurate and reasonable, and that, outside religious spheres, the authority of the Santos Padres (Holy Fathers), over historical matters, might be questioned if they have depended on the findings of writers who had no personal knowledge of the subjects concerned.[14] Accordingly, after sifting arguments, including those on style, for and against the authenticity of letters supposedly written to each other by St Paul and Seneca, he agrees with Erasmus that the evidence is unconvincing.[15] Segura, then, was capable of examining attractive theories with objectivity. But when such enlightened principles and arguments are buried in pedantic verbiage, a plethora of Latin quotations, and superficial recourse to outside reference, it is no wonder, to the present-day reader, that the *Diario de los literatos* should ignore Segura's inconspicuously professed ideal of critical courtesy. More significantly, just as Torres had been able to conceive of the nature of scientific evidence in terms of astrological assumptions, so Segura's idea of the constitution of evidence and proof was unclinical. He understood new trends of enlightenment in part and in part could respond. What he could not understand was the *Diario*'s seeming creation of a Faculty of Criticism: a force of alarm, because of its tone of constituted superiority. Wholesale condemnation of the *Norte crítico* certainly did not encourage development among the Segura-*vulgo* of the incidental signs of enlightened reasoning. Shock-tactics have a revolutionary value in that they can penetrate to fundamentals, consequently predicting warfare, and so promoting general alarm. The *Diario* did not usually err on the side of caution or discrimination and concentrated its 'harsh' judgment on the *Norte crítico*'s all too obvious defects. Besides, Segura had criticized Feijoo, whose open mind was too open for Segura's sense of intellectual safety, but whom the *Diario* recognized as a colleague of driving force. Certainly the *Diario*'s sarcasm in its merciless inspection of every component, major or minor, of the *Norte crítico*, with some personal references – for example, to Segura's supposedly decrepit age – did not encourage Segura's reasonableness to develop at its best. His reply, the 'Apología contra los *Diarios de los literatos*', of 1738, is therefore an attack against the *Diario*'s professed impartiality and erudition, its self-importance, *feijonismo* and inefficiency:

> . . .ellos [the critics in the *Diario*] dan testimonio de sí mismos de que son capaces, hábiles, y de . . . vastísima erudición, que requiere el ejercicio de Censores Generales de los libros . . .
>
> (. . . they assert about themselves, that they are capable, clever and . . . of the enormous erudition required in the exercise of General Judges of books . . .)

They assert that they will proceed with absolute impartiality:

> En el *Diario* se encuentra muy al contrario de todo lo referido . . . porque en su obra hay lisonjas y mordacidad, disímulas notables, y maledicencia muy injuriosa.[16]
>
> (In the Journal one finds the opposite [tendency] to all that it says . . . because in that work there is flattery and mordacity, notable dissimulation, and very injurious slander.)

– this together with a malicious reference to the fact that the *Diario* had not been appearing regularly, and his assumption thereby that the periodical had lost, or had never gained importance.

Certainly in this case neither Segura nor the *Diario* had served the cause of enlightenment with much credit. From the standpoint of semi-*ilustrismo* Segura could hardly be expected to appreciate the *Diario*'s historical significance as an experiment in professionally enlightened criticism. Yet he was not so far from the truth in regarding its technique as inferior to that of, for example, the *Mémoires* of Trévoux. The *Diario* was still in the process of trying to practise a new profession, so was still an instrument of discomfiture. Processes are difficult for minds outside *vulgo*-tradition to categorize.

Scholars more generally worthy of enlightened consideration than Ozejos or Seguras, also found reason for concern in the *Diario*'s tone and technique. A 'Don Plácido Veranio', professedly a 'friend' of the eminent Bibliotecario del Rey (Royal Librarian), Gregorio Mayáns, but in fact Mayáns under pseudonym, voices several reasons for opposing the *Diario*'s treatment of his *Los orígenes de la lengua española*, and reprehends his 'falsos críticos' (false critics) for their 'presunción de saber' (presumptions of knowledge), their trick-sophistry, envy, terminological inaccuracies, and tendency to take Mayáns' statements out of their context etc.[17] Apart, however, from matters of personal pride – and Mayáns admitted to certain 'unimportant' inaccuracies of his own[18] – some of his objections to the *Diario*'s character reflect wider criteria. Understandably enough at a period before the development of multiple learned journals, each with its own service of professional reviewers related to its specific Academic Discipline, the knowledge of critics in such fledgling journals as the *Diario* would of necessity be general and so disruptive. Even Feijoo and his compeers, for all the soundness of their general scholarship, could, by publicizing their judgment on matters outside their own Faculties, create understandable confusion and opposition.

More particularly one notices Mayáns' expression of peninsular self-consciousness and national vigilance, comparable with expressions of British insularity, when dealing with foreign influence as publicized by, for instance, foreign journals. Not that he objected on principle to foreign journals as such. What caused him concern was, he said, the *Diario*'s tendency, as an effect of its authors' foreign focus, to 'quitar la estimación a los escritores de España' (to deprive Spanish writers of esteem):[19] an understandably human and serious concern of national individuality which only time and greater experience could dissipate.

Reasonable in most ways is the article of Dr Joseph Berni, Abogado de los Reales Consejos (Advocate of the Royal Council), who answers criticisms of the *Diario* against works, written respectively by himself and his brother Juan Bautista, a priest, on Law and Religion. His argument that a certain inaccuracy of authorship made by his brother is a mistake such as is easily made by any writer, and is paralleled by mistakes, which he duly instances, often made by the *Diario*, does him of course no credit. Elsewhere similar parallels of spelling mistakes and faults of style hardly further his cause, though his reference to the Diarists' effort and need to capture general attention at all costs was a realistic comment on the life of an experimental journal. More importantly, in answering the *Diario*'s criticism of his *Abogado instruido* that he 'abunda de noticias . . .' ([over] abounds in information), he protests with seeming reasonableness that the Diarists are misinterpreting legal phraseology, that they do not fully understand the technicalities of the Letter of the Law which he quotes now, in relation to various cases, for their benefit:

> Que Vds. no tengan noticia, no es mucho, pues tampoco la tengo de materias médicas, que no he profesado.[20]
>
> (That you should not be [well] informed [on the subject] is not surprising, for I myself am not well informed on medical matters, since I have not studied such a subject [as Medicine].)

Again, then, it is understandable that, rightly or wrongly, the blatantly professed enlightenment of the *Diario* on any matter, academic or otherwise, should engender reasonable provocation. Newly extended ground for specialized argument was fertile but as yet largely untilled. In 1748 the broadminded *Resurrección del Diario de Madrid* . . . imaginatively summarized the main types of public apprehension, existing timelessly, as follows:

> Los *desengaños*, la *claridad*, y las *verdades*, siempre fueron Personajes mal recibidos del Público. Esto se compone facilísimamente con un nuevo bautismo de voces a contemplación de cada uno. A los desengaños se llaman locuras, o novedades peligrosas. A la *claridad*, insolencia. A las *Verdades*, atrevimientos, llenos de reprehensión. Y, a este paso, no hay cosa ya en el Theatro del Mundo, que no se trastrueque, y equivoque. Mutaciones de Comedias parecen los juicios de los hombres. Tramoyas son la variedad de los dictámenes. Por esto no se atreve a salir la verdad en público.[21]
>
> (*Disillusion*, *clarity*, *truth* were always characters badly received by the public. This is easily remedied by baptising each one concerned with a new name. *Disillusion* is called madness, or dangerous novelty. *Clarity* is called insolence. *Truth* is called reprehensible daring. And at this stage, there is nothing now in the Theatre of the World which is not transformed and equivocated. The judgments of men seem like the Dramas' changes of scenery. Varieties of opinion are like stage-machines. For this reason truth never dares to appear in public.)

Yet even this enlightened critic, who had applauded Bacon, Martínez, Feijoo, and the like, and with vitriolic energy had derided Torres' faulty conception of demonstrative evidence in Astronomy,

Medicine and Science, even he is found to be parading a certain apprehension of his own. Was it humanly wise, for purposes of intellectual communication, to write in the *vulgar* tongue?:

> Tales materias piden el sello de la Lengua Latina, para que el Vulgo bajo no las maltrate. Constituye éste muy mal, lo que no entiende bien. Y mal puede entender, lo que está tan fuera de su capacidad . . .[22]
>
> (Such material requires the stamp of the Latin language, to prevent the general public from mistreating it. This public puts together very badly what it does not understand well. And it can understand only badly what is so far beyond its comprehension.)

His reservation can be compared with modern reservations about publicity accorded to certain problems of Medicine, Science, Theology or Sociology technically understandable only by those professionally trained to analyse them in the right context. Nowadays such problems would not of course be shrouded in Latin, and apprehensions about publicizing disturbing problems in their full complexity can still prevail. Those who would hesitate to discuss them outside their professional contexts are being, not necessarily perverse, but merely thoughtful. They are cautious scholars who inhabit grey areas of passage where, as would-be rationalists, they are trained in practical research to establish indisputable proofs of their instinctive enlightenment.

NOTES

1 Francisco Aguilar Piñal, *La Prensa española en el siglo XVIII . . .* (Madrid: C.S.I.C.,1978); Henry F. Schulte, *The Spanish Press, 1470-1966* (Urbana/London: University of Illinois Press, 1968), pp. 86 ff.

2 See *Gaceta de Madrid*, 23.III.19; 12.III.20; 12.XII.19; 14.IV.22; 7.VIII.22.

3 *Gaceta de Madrid*, 6.X.22.

4 *Gaceta de Madrid*, 15.X.20; 8.IV.21; 29.VIII.24; 5.IX.24; 15.X.20; 29.IV.21.

5 *Gaceta de Madrid*, 1.VI.28; 15.X.28.

6 *Mercurio literario . . .* (Madrid: Imprenta del Reino, 1739), Vol.I, p.4. An excellent analysis of the *Diario*'s attitudes is given by Dr Jesús M. Ruiz Veintemilla in his Introduction to the new, facsimile edition of the *Diario de los literatos* (Barcelona: Puvill Libros, 1987), 7 vols. See Vol. I, pp. 7-104.

7 *Mercurio literario . . . Ibid.*

8 *El sol de los Anacoretas, La luz de Egypto, El pasmo de la Tebayda, El asombro del mundo, El portento de la Gracia, La milagrosa Vida de San Antonio Abad* (Madrid, 1737).

9 See *Las Impresiones y Plumistas de la Corte, en busca del Diario Apologético de las murmuraciones* (Madrid: Alonso Balvas), 1738, p. 20.

10 See *Diario de los literatos . . .* Vol. IV, Art. XVI, p. 344.

11 See Ozejo's *Impresiones y Plumistas . . ., ed.cit.*, Censura.

12 *Norte crítico*, Valencia, 1737, p. xxiii.

13 *Op.cit., Discursos* I and II, *et passim.*

14 *Op.cit.*, see *Discursos* IV and VII.

15 *Op.cit.*, see *Discurso* V.

16 *Apología contra los Diarios de los literatos de España sobre los artículos XII, XIII, del Tomo II, y I del Tomo III* (Valencia: Joseph Lucas, 1738), pp. 2-4.

17 Don Plácido Veranio, *Conversación sobre el Diario de los literatos de España* (Madrid: Juan de Zúñiga, 1737), p. 1.

18 *Op.cit.*, pp. 8 ff.

19 *Op.cit.*, p. 4.

20 Joseph Berni, *Satisfacción a los artículos primeros del primero y séptimo tomos del Diario de los literatos de España* (Valencia: Joseph García, 1742), p. 16.

21 Resurrección del *Diario de Madrid, o Nuevo Cordón crítico general de España* . . . (Madrid, 1748). See Prólogo.

22 *Op.cit.*, p. 33.

INDEX